ANTENNAS FOR IoT

For a complete listing of titles in the
Antennas and Propagation Library,
turn to the back of this book.

ANTENNAS FOR IoT

Prutha P. Kulkarni

ARTECH
HOUSE

BOSTON | LONDON
artechhouse.com

Library of Congress Cataloging-in-Publication Data
A catalog record for this book is available from the U.S. Library of Congress.

British Library Cataloguing in Publication Data
A catalogue record for this book is available from the British Library.

Cover design by Joi Garron

ISBN 13: 978-1-63081-993-4

© **2023**
Artech House
685 Canton Street
Norwood, MA 02062

CONTENTS

1

INTRODUCTION   1

2

INDUSTRY 4.0   25

3

RF AND MICROWAVE ASPECTS OF IoT   47

4

BASIC ANTENNAS IN IoT   75

5

ANTENNA MEASUREMENT SYSTEMS   109

6

MINIATURIZATION TECHNOLOGIES   131

7

IoT IN SMART CITIES   151

8

IoT IN WIRELESS COMMUNICATION   175

9

IoT IN SURVEILLANCE AND SECURITY   197

10

RELIABILITY AND SECURITY CHALLENGES   223

11

FUTURE SCOPE AND CONCLUSION   247

1

INTRODUCTION

Internet everywhere and Internet of Things (IoT) everywhere! This timeline and the future ones are into connecting with everyone, by whatsoever means. Communication from tip to toe digitally says it all. Technology is globally advancing tremendously with a smarter speed far more than the earlier human races ever dreamt of. The connectivity option, IoT, incorporates several heterogeneous end systems, such as smart homes, the industrial IoT, smart agriculture, and smart cities, and the smart grid connects them all. Building a wireless product involves more unknowns than just market fit and return on investment. Some unknowns are more grounded in physics, but they still affect the economics of a product. Range is one such unknown. But what is range? Let's put it simply: How far apart can we place two devices and still have working communication between them? Why does the wireless range matter? Because if we exceed the wireless range, the network quality will suffer. We want to avoid data loss, late delivery of data, a reduction in the number of devices we can support, and potentially even losing control of our devices because they are out of range. If we know what range to expect, we can answer questions like: How far apart can you place devices? How many devices can you have in a network? How much data can you send in the network? What will the installation procedures look like? These questions are highly important from a business perspective. How important? In one instance, wireless range led up to a $430 million lawsuit. Theoretical predictions: What are they? Let's first look at what the theory tells us.

This theory was developed in the early 1940s and is called the Friis equation. Loosely put, the Friis equation predicts the transmission range for a given transmission power. Wireless chip producer Texas Instruments published a range estimator tool (swsc002a) to make it easier for people using their chips to use the Friis equation to predict the wireless range. It is a great way to see the theory in action. Set the approximate conditions of two devices, such as the antenna used and data rate, and it will give an estimated range based on the Friis transmission formula. There is an exhaustingly large number of factors coming into play, and rarely the surface is scratched here. It's impossible to account for everything at this stage. And this is the point—any such tool will ultimately lack simply due to all the unknowns. But not so fast! In the real world, radio waves are reflected, attenuated, and interfered with, and the conditions will be different for each and every device that is put out there. Even moving a few centimeters can mean the difference of being heard and not being heard. The theoretical models do not account for these real-world effects.

1.1   ANTENNA APPLICATION AREAS IN IoT

The IoT and smart industrial applications have led to explosive growth, creating scientific and engineering challenges in the development of antenna systems that are efficient, cost-effective, scalable, and reliable. Printed antennas are preferred in portable IoT applications, especially in 5G, while radio frequency identification (RFID) antennas with tuning capabilities are also popular in IoT-enabled environments [1].

To achieve high spectral efficiency, high data rates, and extreme spatial multiplexing of densely distributed users, it is essential to exploit the large number of degrees of freedom (DoFs) of massive multiple input/multiple output (MIMO). Enhanced mobile broadband (eMBB) is a natural evolution of Long-Term Evolution (LTE), with the primary goal of increasing user data rates and network spectral efficiency. The enhancements provided by eMBB primarily target human-type traffic, such as high-speed, wireless broadband access, ultrahigh-quality video streaming, virtual reality, and augmented reality.

In contrast, ultrareliable low-latency communication (URLLC) and massive machine-type communication (mMTC) are essential for machine-type traffic, enabling various types of IoT connectivity. URLLC, also known as mission-critical IoT, envisions transmission of moderately small data packets with extremely high reliability. Advanced antenna systems (AAS) are becoming more widely used, comprising an antenna array closely integrated with hardware and software to implement features such as steerability for adapting antenna radiation patterns to rapidly time-varying traffic and multipath radio transmission conditions.

Choosing an antenna for IoT applications is a complex process, with up to 12 antennas required in a mobile phone–sized device. These antennas must manage different redundancies and services while radiating and receiving independently from one another, putting a premium on antenna isolation. Surface mount antennas in IoT devices are efficient and offer repeatable manufacturing. Maximizing the ground plane and improving battery life help in granular data tracking and avoiding interference among the radios involved.

The selection of an antenna system is an important component of all these node end smart devices, in addition to the complex communication protocols. One of the biggest design challenges is deciding which antenna is best for a given application. Making room for antennas is getting more and more difficult as IoT modules are getting smaller and incorporate more wireless technologies. In order to retain adequate antenna performance under challenging circumstances like noise, fading, and the need for efficiency, IoT-module antenna design must contend with the limitations of ever-shrinking footprints. For the development of effective antenna systems for IoT, improved multiplexing, interference reduction, scheduling, and radio resource allocation approaches collaborate with the antenna design. The main drive of the creative research effort in development of effective, affordable, scalable, and reliable antenna systems is the explosive growth of the IoT and smart industrial applications creating many scientific and engineering challenges.

A rising number of antenna options are available to IoT manufacturers in this sector, which is highly competitive. Prior to the IoT, antennas were essentially an afterthought, but as interoperability and machine-to-machine connectivity have advanced, as well as consumer desire for a seamless smart experience, antenna stock has increased within the IoT ecosystem. Each antenna type has advantages and disadvantages, just like most product lines. Because they depend on their operational environment, antennas are known for having rigid designs. As a result, there are a few things to take into account during the design and production procedures. The complex balancing act between physical size, performance, and cost that antenna designers frequently face is greatly influenced by a variety of factors and IoT specifics.

1.2 SENSORS AND ANTENNAS

In the IoT, sensors [2] are deployed at dispersed nodes and then link together via networked computing devices to transmit data for further analysis and storage. In 2015, 25 billion sensor gadgets were connected to the internet, providing a glimpse into the future. Cisco's Internet Business Solutions Group (IBSG) estimated approximately 50 billion devices to be interconnected for the year 2020. As more and more gadgets become internet-connected, it becomes increasingly important that each individual node device be equipped with some sort of communication system. Depending on the type of sensor being used, its physical dimensions will vary. As a result, smaller antennas must be incorporated beside the sensor without burdening the device. Therefore, custom antennas are developed for each sensor or device. In most industrial, environmental, and logistical applications, different sensor nodes are placed at remotely large distances, making it difficult and impractical to power them using batteries or other sources. The antenna designs face challenges like miniaturization, compactness, and low power as more sensors and circuit devices are connected using IoT. Many sensors can function on very little energy. Because sensors are surrounded by multiple RF sources like Global System for Mobile Communications (GSM)/LTE, Wi-Fi, WiMAX, wireless local area network (WLAN), and others, radio frequency (RF) energy harvesting can be used to produce low power supply to sensors. Many different types of actuators, micro-electromechanical systems (MEMS) devices, batteries, cells, energy-harvesting methods, radio modules, and antennas are used in the many different end devices that make up the IoT ecosystem. Rectification is not achievable in the RF harvesting system, where RF power is harvested via antennas, then sent to rectifying circuits, followed by power combiners, and finally delivered to load. Antennas are custom-tailored for each wireless application in terms of size, transmission power, and frequency range. However, there are still major distinctions between the many available wireless network topologies.

1.2.1 IoT Topologies

IoT topologies refer to the network architecture and connectivity arrangements used in IoT systems. The selection of a particular topology depends on the specific application requirements, available resources, and other factors such as scalability, reliability, and cost.

Here are some of the common IoT topologies:

1. *Mesh topology:* In this topology, devices are interconnected through multiple wireless links forming a network mesh. Each node in the network can transmit and receive data, and the mesh architecture provides redundancy, scalability, and fault tolerance.

2. *Star topology:* In a star topology, IoT devices are connected to a central hub or gateway. All data flows through this central point, which makes it easier to manage and monitor the network. This topology is simple to set up and maintain, but it can be less fault-tolerant than other topologies.

3. *Bus topology:* In a bus topology, IoT devices are connected to a single communication line. Data is transmitted serially on this line, and each device reads the data intended for it. This topology is relatively easy to implement, but if the communication line fails, the entire network can be affected.

4. *Tree topology:* The tree topology is similar to the bus topology, but it has a hierarchical structure. IoT devices are connected to a central hub or gateway that is then connected to other hubs or gateways forming a tree-like structure. This topology is useful for large-scale IoT systems that require scalability and fault tolerance.

5. *Hybrid topology:* A hybrid topology combines two or more topologies to form a network that meets specific requirements. For example, a mesh-star hybrid topology can provide the redundancy of a mesh network while still maintaining the central control of a star network.

Overall, the choice of IoT topology depends on the specific application requirements and the available resources. Each topology has its advantages and disadvantages, and selecting the right one requires careful consideration of these factors.

1.3 IoT IN ACTION

It's no secret that IoT will explode in popularity over the next few decades. Short-range IoT devices [3] are expected to outnumber mobile phones, which means IoT will soon pervade every sector of the economy, from manufacturing to retail to the home. While specialized proprietary hardware and a lack of vendor neutrality provide challenges, the ecosystem of IoT end devices, gateways, modems, and base stations is in rife with standard bodies and protocols designed to address these issues. IoT is already in action in many fields and is transforming the way businesses operate and people live their lives. Here are a few examples of IoT in action:

1. *Smart homes:* IoT devices such as smart thermostats, smart locks, and smart lighting systems are becoming increasingly popular in homes. These devices can be controlled remotely using a smartphone, and they can also learn user preferences and adapt to them automatically.

2. *Industrial automation:* IoT sensors and devices are being used to automate industrial processes, making them more efficient and reducing the risk of accidents. For example, IoT devices can monitor equipment performance, detect faults, and trigger maintenance alerts before a breakdown occurs.

3. *Healthcare:* IoT devices are being used in healthcare to monitor patient health remotely. Wearable devices such as fitness trackers and smartwatches can track vital signs such as heart rate, blood pressure, and glucose levels, enabling healthcare professionals to provide more personalized care.

4. *Smart cities:* IoT sensors and devices are being used to manage city infrastructure more efficiently. For example, smart traffic lights can adjust their timings based on real-time traffic flow, reducing congestion and improving air quality.

5. *Agriculture:* IoT sensors and devices are being used in agriculture to optimize crop yields and reduce water usage. Sensors can monitor soil moisture levels, temperature, and humidity, enabling farmers to make data-driven decisions about irrigation and fertilization.

These are just a few examples of how IoT is transforming various industries. As IoT technology continues to advance, it is likely to create even more opportunities for innovation and disruption in the years to come.

1.4   INDUSTRIAL IoT

While there is no universal IoT platform, partnerships work to optimize hardware and software for specific use cases, increasing the likelihood that a given use case can make use of a standard physical (PHY) and medium access control (MAC) layer. For instance, the HART Communication Foundation has established the WirelessHART standard for the industrial IoT; it features medium speed (150 Mbps), low latency (1 millisecond), excellent reliability, and decent battery life (3 to 5 years) for Industrial IoT (IIoT). For applications in agriculture, industry, medicine, and smart cities, the emerging low-power wide area network (LPWAN) designs permit ultrahigh connection distances (1 km), extremely high battery lives (10 years), and low throughput on the scale of bit per second (bps) transmissions. Low bandwidth uses have unique antenna requirements.

The IoT is unlike extremely high–throughput wireless networks such as 5G, high-efficiency wireless (HEW)—also known as IEEE 802.11ax—and WiGig that boost speeds on the order of 10 Gbps by utilizing vast continuous spectrum space in the millimeter-wave (mmWave) bands or by using high order advanced modulation schemes like carrier aggregation, 64-QAM orthogonal frequency-division multiplexing (OFDM), and multiuser MIMO (MU-MIMO). In massive MIMO and microcell deployments, networks like these frequently employ state-of-the-art antenna configurations like active electronically scanning arrays (AESAs), which can take the form of phased array antennas or switching beam arrays.

In contrast to LTE-A, which can use up to 20 MHz of bandwidth with carrier aggregation and beyond with mmWave spectrum use for 5G, IoT networks and devices make use of the licensed and unlicensed sub-6 GHz bands using very little bandwidth (5 MHz). Instead of relying on beamforming algorithms with AESAs to intelligently direct uplinks and downlinks between gateways and end devices, IoT networks make use of star, mesh, or point-to-point topologies as seen in Figure 1.1.

This usually necessitates the employment of very basic omnidirectional antenna structures, such as chip, printed circuit board (PCB), whip, rubber duck, patch, and wire antennas, to carry out the task of establishing a link. Radio modules and development kits for the IoT are widely available, and many of them include Global Positioning Systems (GPS), Bluetooth, and Wi-Fi antennas. Examples include Qualcomm's Internet of Everything (IoE) development platform and the Arduino GSM.

Communication between nodes and between gateways and nodes in the IoT is accomplished through the use of predefined topologies, most frequently star and mesh. Required link distance (mesh topologies are restricted to very short distances), battery usage (mesh must always be on standby while star-based devices can enter sleep modes), and latency determine the optimal network structure (depending up on the number of hops from nodes to get to the cloud).

1.4.1   Details of Applications in IoT with Frequencies

IoT applications use a wide range of frequencies depending on the type of device and the application requirements. Here are a few examples of IoT applications and the frequencies they typically use:

1. *Smart home devices:* Many smart home devices, such as smart thermostats, security cameras, and smart lighting systems,

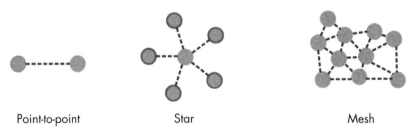

Point-to-point   Star   Mesh

Figure 1.1   IoT network communication topologies.

use Wi-Fi frequencies (2.4 GHz and 5 GHz) for connectivity. However, some devices may also use other frequencies such as Bluetooth (2.4 GHz) or Zigbee (2.4 GHz).

2. *Industrial automation:* IoT devices used in industrial automation applications often use sub-GHz frequencies such as 900 MHz or 2.4 GHz to avoid interference from other wireless devices in the environment.

3. *Healthcare devices:* IoT devices used in healthcare applications often use Bluetooth (2.4 GHz) or Zigbee (2.4 GHz) for connectivity. However, some devices may also use cellular frequencies such as LTE-M (Cat-M1) or NB-IoT for remote monitoring applications.

4. *Smart cities:* IoT devices used in smart city applications may use a range of frequencies depending on the specific application. For example, smart traffic lights may use 900-MHz frequencies for long-range communication, while smart parking systems may use Bluetooth (2.4 GHz) for short-range communication.

5. *Agriculture:* IoT devices used in agriculture applications may use a range of frequencies depending on the specific application. For example, soil moisture sensors may use sub-GHz frequencies to achieve longer range and better penetration through soil, while drones used for crop monitoring may use Wi-Fi (2.4 GHz or 5 GHz) frequencies for remote control and data transmission.

Overall, the frequency used by an IoT device depends on the specific application requirements such as range, data rate, and power consumption. As IoT technology continues to evolve, it is likely that new frequencies and wireless protocols will emerge to meet the needs of emerging IoT applications.

1.5 ANTENNA PARAMETERS

Antennas are essential components in any wireless communication system. They play a critical role in transmitting and receiving electromagnetic signals, and their characteristics determine the quality of wireless communication. Here are some of the most important characteristics of antennas [4]:

1. *Frequency range:* The frequency range is the range of frequencies over which an antenna is designed to operate. Antennas are designed to work within specific frequency bands, such as ultrahigh frequency (UHF), very high frequency (VHF), or GHz, depending on the application.

2. *Gain:* Gain is a measure of the ability of an antenna to direct and concentrate energy in a specific direction. It is usually measured in decibels (dB) and is relative to an isotropic radiator (an ideal antenna that radiates energy uniformly in all directions).

3. *Directivity:* Directivity is a measure of the ability of an antenna to concentrate energy in a particular direction. It is the ratio of the maximum radiation intensity in a specific direction to the average radiation intensity over all directions.

4. *Radiation pattern:* The radiation pattern is a graphical representation of the directionality of an antenna's radiation. It shows the distribution of energy in space around the antenna.

5. *Polarization:* Polarization is the orientation of the electric field of the electromagnetic wave with respect to the antenna. It can be either linear or circular, and different types of antennas are designed to work with different polarizations.

6. *Impedance:* Impedance is the measure of opposition to the flow of an alternating current in an electrical circuit. The impedance of an antenna must match the impedance of the transmission line to ensure maximum power transfer and efficient operation.

7. *Bandwidth:* Bandwidth is the range of frequencies over which an antenna can operate without significant loss of performance. The broader the bandwidth, the better the antenna is at receiving and transmitting.

Gain and directivity are the two main characteristics that characterize antennas. The directionality of a beam of radiation describes how strongly it is focused in one direction. Therefore, directional antennas have smaller radiation patterns than omnidirectional ones, while omnidirectional antennas are rather uniformly focused in all three dimensions. Figure 1.2 shows the directional antenna patterns with specific antenna gains.

This is typically achieved by mixing several different types of radiating pieces. The gain of an antenna is a common metric listed in technical documents, and it is used to rate the amount of power emitted by the antenna.

For the IoT to operate in the unlicensed industrial, scientific, and medical (ISM) bands, the equivalent (or effective) isotropic radiated power (EIRP) must be less than or equal to that of a licensed digital transmitter, reducing the potential for interference in the increasingly crowded radio spectrum. To achieve the same signal intensity as the antenna under test (AUT) in the direction of its strongest beam, the ideal isotropic antenna would have to emit a total power equal to the EIRP. This metric gives a more complete picture of the transceiver module because it considers not only gain but also transmitter output power and antenna feed loss.

1.6   ANTENNAS USED IN IoT DEVICES

Wire, whip, rubber duck, paddle, chip, and PCB antennas are all examples of common types of IoT antennas [5]. For uses that call for the

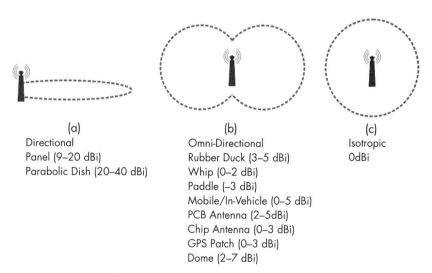

(a)
Directional
Panel (9–20 dBi)
Parabolic Dish (20–40 dBi)

(b)
Omni-Directional
Rubber Duck (3–5 dBi)
Whip (0–2 dBi)
Paddle (–3 dBi)
Mobile/In-Vehicle (0–5 dBi)
PCB Antenna (2–5dBi)
Chip Antenna (0–3 dBi)
GPS Patch (0–3 dBi)
Dome (2–7 dBi)

(c)
Isotropic
0dBi

Figure 1.2   Patterns of radiating antenna depending on the type of gain.

antenna to be routed along the device's body, there are variations with dual/multiband antennas and flexible antenna constructions (e.g., unmanned aerial vehicles (UAVs)). Prior to this, it was stated that the frequency of the specific IoT application must be within the antenna's bandwidth. Several typical IoT applications are shown in Figure 1.3 together with the wireless networking technologies they employ and the frequency ranges in which they operate. Despite the fact that most of these use cases operate in the unlicensed ISM bands, special frequencies are reserved for things like medical body area networks (MBAN) and wireless avionics intracommunications.

It's important to think about both bandwidth and physical space. Due to the fact that they are not soldered directly onto the PCB of the IoT device, whip, rubber duck, and paddle antennas are more easily removable, making them ideal for use in early prototypes. The quarter-wave whip is the most popular form of monopole antenna. Both sides of a dipole antenna meet at the feed point, as do both parts of a monopole antenna, but in the latter case the ground plane takes the role of the second radiating element. When using a coaxial cable to power a monopole antenna, such as a whip antenna, the shield is often used as the ground plane. In most cases, quarter-wavelength radials positioned perpendicular to the antenna will be used in large monopole antenna systems. The radiation pattern of a center-fed dipole is created when waves are reflected by an ideal ground plane so that the rear side of the ground plane seems to have another identical monopole antenna. The virtual image antenna cannot be realized with a real ground plane, although a sufficiently conducting ground plane is still required. Radiation efficiency (inversely proportional to ground plane resistance), impedance, resonant frequency, and performance might all suffer if these conditions are not met. The monopole antenna's smaller angle of radiation allows for greater propagation distance, especially at low UHF frequencies, when compared to the more common dipole antenna.

Whip antennas are also available with ingress protection (IP) ratings, such as IP67 or IP68, depending on the design. This is especially crucial in tough industrial or outdoor settings where water, dirt, oil, and chemicals can quickly destroy external antennas.

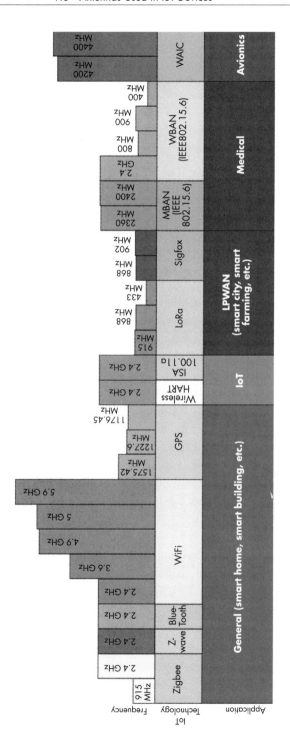

Figure 1.3 Operational frequencies in IoT technologies.

1.6.1   Considerations for Patch Antennas

Patch antennas [6] are frequently used in IoT devices that have GPS capabilities. This is due to the fact that signals transmitted by satellites are frequently either desioned for right-handed circular polarization (RHCP), or left-handed circular polarization (LHCP), while patch antennas can be designed for dual polarization. This dual polarization is frequently reconfigured by switching a perturbation element using PIN diodes or RF MEMS devices. Having said that, there are still numerous patch antennas that will only exhibit a single form of polarization, either linear, RHCP, or LHCP; hence, it is extremely important to select a polarization that is compatible with the transmission.

1.6.2   Chip and PCB Antenna Considerations

By utilizing IoT devices with embedded antennas, such as chips and PCBs, the size of a sensor node can be reduced, resulting in a more compact device. Due to the presence of conductive traces on the board, PCB antennas [7] typically have higher gains when compared to their chip-based counterparts. In the case of PCB antennas, some of the possible topologies, or configurations, are the inverted F, the L, and the folded monopole. In the process of fabricating PCB antennas, the ground plane is an extremely important component. A ground plane that is too small can severely limit design freedom because of the negative consequences it has on functional bandwidth, antenna radiation efficiency, and radiation pattern. Antennas that are embedded in PCBs typically take up a sizable fraction of the available space on the board. The gain of the antenna is proportional to the volume of the radiative element; hence this is necessary. PCB antenna–using sensor nodes, on the other hand, have the benefit of keeping relatively flat, which may make it easier to encapsulate and install them anywhere.

Since omnidirectional chip antennas feature a cardioid radiation pattern, it should not come as a surprise that these antennas have the lowest gains. When tested in real-world situations, it was discovered that these antennas have greater directionality in a particular direction, which indicates that they may have a more ideal orientation for link formation. Chip antennas are ideally suited for use in wearable IoT applications such as wireless body area networks (WBANs) due to the tiny benefits they offer together with an acceptable trade-off in the form of shorter communication distances.

1.6.3   IoT Directional Antennas

The range of a transmission can be increased by using a directional antenna [8], such as a Yagi or a sector antenna, or a base station for the IoT. A higher link distance is likely to be achieved with a high-gain directional antenna that can be mechanically shifted than with an omnidirectional antenna configuration. For instance, Yagi antennas are commonly employed in IIoT SCADA (supervisory control and data acquisition) systems because of their high data throughputs and dependability. Panel (sector) antennas, which may be installed in multiples to provide 360-degree coverage, are commonly used in SigFox, LoRa, and WLAN base stations due to their greater beamwidth compared to Yagi antennas.

1.6.4   Trace Antennas

Trace antennas are one of the most affordable solutions in the antenna market, which contributes to their widespread use. They frequently exhibit good performance on single-band platforms and are frequently already placed on PCBs. Additionally, they can be entirely designed and implemented by yourself with little to no experience.

In contrast to other antenna designs, their size implies that they take up a lot of space, and shrinking them will compromise performance. Trace antennas can also have performance problems, particularly with portable and wearable electronics. RF difficulties with internal and external trace antennas are also possible, especially in dynamic situations.

1.6.5   Flexible Printed Circuit

There are various IoT technological fields that use this bendable antenna, including wearables, telecoms, and automotive fields. Flexible printed circuit (FPC) [9] antennas have an advantage over their rigid equivalents, particularly during research and development, because of their adaptability and small size. As an added bonus, the flexible film allows for installation on uneven surfaces, and it may be attached via wire to compact IoT devices with limited PCB real estate. In addition to being lightweight and inexpensive, FPCs also have a low price. Connecting FPC antennas to IoT devices and PCBs is a breeze thanks to the antennas' compatibility with a wide range of connections, including underground feeder (UF) cables.

When not handled with care, FPCs might be damaged. There is a high degree of difficulty in making changes or repairs, and extra care must be used during installation to ensure there is no interference with other parts. To maximize performance, FPC antennas are kept at least 10 mm away from anything metallic. Many IoT manufacturers are not yet equipped to employ FPCs, which contributes to interoperability problems.

1.6.6 Wired Helix

Inexpensive and straightforward, a wired helix antenna [10] is a monopole design that twists a single wire. Their spiral design makes them ideal for tight quarters and makes them easy to reach for anyone. The lower frequency bands (high frequency (HF), VHF, and UHF) are where wired helix antennas shine, making them ideal for use in portable communications devices. They are reliable and provide excellent focus.

However, these antennas can only be used with basic IoT gadgets that operate on a single radio or frequency band. Due to the fact that frequency drops down with increasing size and turn count, designers face a variety of obstacles.

1.6.7 Surface Mount Technology Antennas

Surface mount technology (SMT) antennas [11] and compact antenna designs gained popularity with the proliferation of handheld electronics. These tiny antennas are usually attached to a PCB via solder. They are simple to include with pick-and-place technology, produce little to no noise, and weigh very little. IoT designers can free up more PCB real estate because of the compact nature of SMT antennas. The antennas are able to integrate various frequencies in a solid, embedded form. In most cases, the board costs, material handling costs, and regulated manufacturing processes associated with producing SMT antennas can be reduced. To function properly, SMT antennas need a large ground plane at low frequencies and a clearance area, both of which increase the amount of space needed for installation. Antenna optimization requires careful consideration of PCB placement and the resulting clearance space. Lastly, cost is proportional to both square footage and the range of frequencies needed.

1.6.8 3-D Antennas

Advanced technical techniques, like laser direct structuring (LDS), allow these antennas to be developed and manufactured on a 3-D plastic carrier [12] or mold. The 3-D design permits engineers to alter and fine-tune the antenna to the client's specifications during the prototype phase. In addition, the client can anticipate speedy design and turnaround. These antennas are an excellent solution when space is limited. In addition, 3-D printer antennas are SMT-compatible and have a low failure rate, enabling continuous mass production. Finally, antenna traces can be altered without affecting the plastic molding, allowing for the incorporation of many frequencies. As 3-D printed antennas are customized items, it can be difficult to obtain certain colors and materials. Consequently, it is crucial to consider and implement 3-D antenna fabrication at the outset of the IoT project.

1.6.9 Smart Antennas

Smart antennas, also known as intelligent adaptive arrays of numerous antennas, employ liquid crystals to dynamically reshape their internal configuration and sophisticated algorithms to determine the most effective antenna configuration. Because of the improved signal-to-interference ratio (SIR), the user experience is streamlined. These intelligent antennas' robust directional capabilities enable a safe, reliable service that performs very well even in extremely crowded areas. In addition, smart antennas' spatial detecting skills make them ideal for geolocation. You can put them in quite different places without worrying about their ability to communicate with one another. They offer the user an extra layer of protection because they are difficult to hack and modify.

Beamforming is a communication method used by smart antennas, and it necessitates the inclusion of a digital signal processor, which can be quite pricey. The transceivers in this antenna are more sophisticated than in other antenna types, so they need to be calibrated often and precisely.

Internal noise and external obstructions are two more factors to think about while designing an antenna.

It is possible for a number of elements, not merely those connected to the topology of the antenna, to obstruct the transmission of a signal. It is conceivable for certain onboard components to generate false signals, which, in the absence of such generators, would interfere with the sending and receiving of signals. This needs to be thoroughly explored, either through simulation software or experiment, so that interference can be reduced in the RF front end. In addition, in order for signals to reach a gateway or end node when propagation is hindered by obstructions such as metal housing, there is a requirement for the use of external antennas. Uplinks and downlinks are susceptible to interference from fading and other multipath effects caused by environmental barriers such as buildings and mountains. Having said that, it's possible that this is only a concern for really dispersed networks. Experimenting with multiple positions for components, antennas, and nodes, as well as different materials for the housing itself, is the most effective way to construct an IoT system. This is also the best strategy to use.

1.7   THE COMPLEXITY OF THE ISSUE

In the same way that there is a large variety of standards and protocols for the IoT, there is also a vast variety of shapes and sizes of antennas that can help broadcast the many different signals that are utilized in the devices that make up the IoT. In order to maintain the safety of industrial wireless sensor networks (IWSN), it is possible that high availability sensor nodes (and antennas) would be required. These nodes must be able to receive firmware updates wirelessly. External antennas with an IP rating will be required for usage in applications in smart farming that make use of a WLAN or a LPWAN. For the vast majority of MBAN applications, it is highly likely that wearable modules with chip antennas will be required. These modules will need to satisfy extremely rigorous power criteria and ultimately interact with a controlling device through a wireless link. Antennas designed for use in smart home and smart appliance applications may not necessarily need to be as power-efficient as antennas designed for use in mobile devices; nonetheless, they still need to be small enough to be contained within a housing. A range of antenna structures are able to provide the most effective accommodation for the transmission characteristics of every particular IoT application.

1.8   WAYS THE ANTENNA IS DECEIVING YOU

The antenna must be adjusted so that it operates in the desired frequency range. The sub-GHz unlicensed ISM bands are normally first
choice, but other bands are considered depending on the use case and
laws (ISM). However, there are regional variations in the sub-GHz ISM
band. The precise frequencies are region-specific. The United States
operates in the 902- to 928-MHz spectrum as defined by Federal Communications Commission (FCC) Code of Federal Regulations (CFR)
15.247, whereas the European Union operates in the 863- to 870-MHz
range as defined by the European Telecommunications Standards Institute European Standards (ETSI EN) 300.220. Now, in order for an
antenna to effectively emit and absorb energy, it is designed for a particular frequency band. Similar to a tuning fork used to fine-tune a
piano, the size of the object determines the frequency at which it resonates. Like a radio receiver, an antenna can be tuned by adjusting its
length. To be multiband, an antenna must be able to operate over a
wide range of frequencies rather than just one. The trade-off is typically lower overall antenna performance compared to a single-band
antenna. When it comes to tuning, voltage standing wave ratio (VSWR)
and bandwidth are the two most crucial measurements. The value of
the VSWR indicates how effectively an antenna is tuned to the source
or transmission line. A measure of the amount of energy that is not
emitted by the antenna but rather reflected back into the transmitter.
As a rule of thumb, a ratio of 2:1 is ideal, but anything higher than
4:1 should be tuned. The range of frequencies over which the antenna
maintains a VSWR of 2:1 is called its bandwidth. Generally speaking, it
is always unclear to notify in which frequency range an antenna is designed to work in just by looking at it; instead, the antenna's datasheet
should be consulted. In other words, use of any old antenna without
knowing exactly the frequency range of operation in IoT is fruitless.

1.9   THE INTENDED USE IS WITH A DIFFERENT ELECTRONICS UNIT

There are specifics that need to be met by the equipment to which
each antenna is affixed. The radiator and the low-impedance return
channel to ground (sometimes called the counterpoise) are the two
fundamental components of the antenna. The frequency being used
determines the length of the radiator. A whole wavelength at 915 MHz

is roughly 33 cm; therefore an antenna operating at that frequency would need to be quite huge. The relationship between frequency and wavelength is straightforward: the lower the frequency, the longer the wavelength. A fortunate fact about antennas is that they can be segmented into half-, third-, and quarter-wavelength diameters. The radiator may be bent into a helix shape, which saves even more room. This is especially frequent with whip antennas and further complicates the task of visually distinguishing between different types of antennas. The complicated phase now begins. Monopole and dipole antennas are those that have and do not use a counterpoise, respectively. When a counterpoise (a dipole antenna) is used with a whip antenna, it is often a radiator mounted on top of a metal pipe. The product itself must serve as the antenna in the case of monopole application, with the metal casing serving as the actual antenna. An extensive ground plane is typically required for chip antennas. Because of this, claiming a small footprint is more complicated than just referring to the size of the chip itself, as the counterpoise board also needs to be considered. A large ground plane, for example, may be necessary for an antenna. If you want to make sure the product meets your needs, the datasheet should be studied carefully. The antenna's characteristic impedance adds another layer of intricacy. In this industry, 50-ohm impedance has become the de facto standard, however 75 and 100 ohm are also used in some niche goods. Therefore, it is important to consider not only the frequency range the antenna is designed to work in, but also its kind (whether monopole or dipole) and, if relevant, its impedance. The enclosure, ground plane, or antenna must all be considered as potential counterpoises.

1.10   VARIATION IN REGULATIONS AND DIRECTIONALITY

They also, predictably, set a cap on the effective radiated power [13], or ERP. This is the maximum allowable amount. Here, mW or dBm are used to indicate transmission power, with 1 mW equating to 0 dBm. The typical upper bound is 25 mW (+14 dBm). Radio waves from antennas rarely spread out in a perfectly circular pattern. The term isotropic is used to describe one kind of perfect antenna. Depending on the antenna's design, real antennas can direct some of that power in a certain way. A donut-shaped distribution of power might arise. Such directionality gain is expressed as a decibel difference (dBi) from the performance of an ideal isotropic antenna. This PCB antenna's radiation pattern is displayed as an example. The design is nearly isotropic in two dimensions and anisotropic in the third, making it easy to see from any angle. Some regulations, like FCC 15.247 for the 915-MHz spectrum, allow average output power (i.e., some directionality is fine). Transmission power must be reduced when dealing with strong directionality, such as a highly anisotropic antenna layout. One must be familiar with the relevant legislation, the radiation pattern around the antenna, and the steps needed to get it approved by regulators. The product's radiation pattern may differ from that stated on some antenna datasheets. In that situation, precise measurements will need to be made by oneself.

1.11 ADDITIONAL ISSUE: THE WRONG PLUG-IN

The FCC in the United States is against having customers swap out their own antennas. Users may introduce interference into the spectrum by employing antennas that aren't designed for the device in question [14]. This necessitates an unconventional, maybe proprietary, or permanently installed connection between the antenna and the product. However, the standard connector with reverse polarity will do the trick.

1.11.1 IoT Sensors Combined with Antennas in a Connected World

IoT sensors and antennas are integral components of a connected world. IoT sensors are used to gather data from various sources, such as temperature, humidity, motion, and light, while antennas are used to transmit and receive data wirelessly. When combined, they enable a wide range of IoT applications, from smart homes to industrial automation.

Antennas are critical to the success of IoT applications because they provide the means for devices to communicate wirelessly over a range of distances. There are many types of antennas used in IoT applications, such as patch antennas, dipole antennas, and Yagi antennas. The choice of antenna depends on factors such as frequency, range, and radiation pattern.

IoT sensors are often combined with antennas to enable wireless communication. For example, in a smart home, temperature and humidity sensors may be combined with Wi-Fi antennas to enable remote monitoring and control of the home environment. In industrial automation, sensors and antennas may be used to monitor and control equipment wirelessly, enabling remote operation and maintenance.

The combination of IoT sensors and antennas can also enable new applications that were not possible before [15]. For example, in agriculture, sensors can be combined with drones equipped with antennas to monitor crops and soil conditions wirelessly. This enables farmers to optimize crop yields and reduce waste by applying fertilizers and pesticides only where they are needed.

Overall, the combination of IoT sensors and antennas is key to the success of IoT applications in a connected world. As IoT technology continues to evolve, we can expect to see new and innovative ways that sensors and antennas are combined to enable new applications and services.

1.12   CONCLUSION

An antenna is a must for each and every IoT gadget. However, as seen, selecting the appropriate antenna is not usually a walkover. One can extend the range and decrease the power needs of the device with a high-quality antenna. In the worst instance, a malfunctioning antenna can make field equipment inaccessible. More and more data transmission is being handled by antennas as the amount of data generated by technology continues to expand exponentially. In addition, the antenna is now a crucial part of any technological solution because of the IoT, smart gadgets, and smart cities. The trend in antenna research seems to be toward higher frequencies, smaller antennas, and more user-friendly layouts that can pick up radio waves from any direction. Larger, less sleek antennas will be replaced by those optimized for the IoT and data transmission in the near future.

By 2023, it is expected that 50 billion gadgets will be linked to the IoT. The interconnection of all things and all people via the internet is rapidly expanding. IoT services and IoT product designs require a much deeper level of design and technology collaboration due to the inherent technological integration and novel consumer experiences inherent to these endeavors. IoT product designs, in contrast to those of previous technologies, necessitate the participation of numerous disciplines, are essential to contemporary enterprises, and have considerable technological appeal. IoT product designs integrate the real and virtual worlds with the underlying network to provide useful data.

References

[1] Vashi, S., J. Ram, J. Modi, S. Verma, and C. Prakash, "Internet of Things (IoT): A Vision, Architectural Elements, and Security Issues," in *2017 International Conference on I-SMAC (IoT in Social, Mobile, Analytics and Cloud) (I-SMAC)*, Palladam, India, 2017, pp. 492–496, doi: 10.1109/I-SMAC.2017.8058399.

[2] Sehrawat, D., and N. S. Gill, "Smart Sensors: Analysis of Different Types of IoT Sensors," in 2019 *3rd International Conference on Trends in Electronics and Informatics (ICOEI)*, Tirunelveli, India, 2019, pp. 523–528, doi: 10.1109/ICOEI.2019.8862778.

[3] Zanella, A., N. Bui, A. Castellani, L. Vangelista, and M. Zorzi, "Internet of Things for Smart Cities," *IEEE Internet of Things Journal*, Vol. 1, No. 1, Feb. 2014, pp. 22–32, doi: 10.1109/JIOT.2014.2306328.

[4] Balanis, C. A., "Fundamental Parameters and Definitions for Antennas," in *Modern Antenna Handbook*, Wiley, 2008, pp. 1–56, doi: 10.1002/9780470294154.ch1.

[5] Tyagi, D., S. Kumar, and R. Kumar, "Multifunctional Antenna Design for Internet of Things Applications," in *2021 7th International Conference on Advanced Computing and Communication Systems (ICACCS)*, Coimbatore, India, 2021, pp. 557–560, doi: 10.1109/ICACCS51430.2021.9441696.

[6] Sharma, M., A. K. Gupta, T. Arora, D. Pandey, and S. Vats, "Comprehensive Analysis of Multiband Microstrip Patch Antennas used in IoT-based Networks," in *2023 10th International Conference on Computing for Sustainable Global Development (INDIACom)*, New Delhi, India, 2023, pp. 1424–1429.

[7] Hietpas, K., "Embedded PCB Antennas for IoT: Design and Implementation Considerations," *Microwave Journal*, Nov. 2021.

[8] Gautam, P. R., A. Verma, S. Kumar, D. Prasad, and A. Kumar, "Design of Directional Antennas for Wireless Sensor Networks and the Internet of Things Experiments," in *IEEE Sensors Letters*, Vol. 6, No. 9, Article No. 6003404, Sept. 2022, pp. 1–4, doi: 10.1109/LSENS.2022.3202919.

[9] Wu, K.-L., G.-Y. Chen, and J.-S. Sun, "Flexible Printed Circuit (FPC) Board for Mobile Antennas Operation," in *2009 4th International Microsystems, Packaging, Assembly and Circuits Technology Conference*, Taipei, Taiwan, 2009, pp. 93–94, doi: 10.1109/IMPACT.2009.5382171.

[10] Lizzi, L., F. Ferrero, P. Monin, C. Danchesi, and S. Boudaud, "Design of Miniature Antennas for IoT Applications," in *2016 IEEE Sixth International*

Conference on Communications and Electronics (ICCE), Ha-Long, Vietnam, 2016, pp. 234–237, doi: 10.1109/CCE.2016.7562642.

[11] Choi, J., J. Park, W. Hwang, and W. Hong, "Millimeter-Wave 5G Antenna-in-Package for Mobile Devices Featuring Intelligent Frequency Correction Using Distributed Surface Mount Technologies," in *2021 15th European Conference on Antennas and Propagation (EuCAP),* Dusseldorf, Germany, 2021, pp. 1–5, doi: 10.23919/EuCAP51087.2021.9410997.

[12] Tsai, M., et al., "Advanced Antenna Integration of 3D System in Package Solutions for IoT and 5G Application," in *2019 22nd European Microelectronics and Packaging Conference & Exhibition (EMPC),* Pisa, Italy, 2019, pp. 1–6, doi: 10.23919/EMPC44848.2019.8951765.

[13] Alabadi, M., A. Habbal, and X. Wei, "Industrial Internet of Things: Requirements, Architecture, Challenges, and Future Research Directions," *IEEE Access,* Vol. 10, 2022, pp. 66374–66400, doi: 10.1109/ACCESS.2022.3185049.

[14] Maeng, S. J., M. A. Deshmukh, G. A. Bhuyan, and H. Dai, "Interference Analysis and Mitigation for Aerial IoT Considering 3D Antenna Patterns," *IEEE Transactions on Vehicular Technology,* Vol. 70, No. 1, Jan. 2021, pp. 490–503, doi: 10.1109/TVT.2020.3046121.

[15] Shafique, K., B. A. Khawaja, F. Sabir, S. Qazi, and M. Mustaqim, "Internet of Things (IoT) for Next-Generation Smart Systems: A Review of Current Challenges, Future Trends and Prospects for Emerging 5G-IoT Scenarios," *IEEE Access,* Vol. 8, 2020, pp. 23022–23040, doi: 10.1109/ACCESS.2020.2970118.

2

INDUSTRY 4.0

2.1 INTRODUCTION

The modern business world has come a very long way since its humble beginnings at the beginning of the industrial revolution in the eighteenth century. The leap from Industry 1.0 to the subsequent industrial age, known as Industry 4.0, has been accomplished in a relatively short amount of time. As production efficiency and scale improved over the timeline for the ever-growing number of customers, businesses have transitioned from having one or two owners and one or two managers, to having a large number of owners, managers, and employees. This was possible because of increased employee numbers. This point marked the beginning of the industrial culture because it was the first time that quality, efficiency, and scale were prioritized from the very beginning of what we now refer to as Industry 1.0. At the turn of the twentieth century, we witnessed the beginning of the second industrial revolution, also referred to as Industry 2.0. The development of machinery that could be powered by electricity was a significant contributor to this change. As a form of power supply, electricity was already being utilized to a significant degree. Electrical equipment was far simpler, easier, and less expensive to operate and maintain in comparison to steam- and water-based machinery, which were inefficient and required a large number of resources. Additionally, during this time period, the first assembly line was created, which considerably eased the production of goods in industrial settings. The assembly line approach of mass production became the industry standard very rapidly. During this time period, the business culture that had its beginnings in Industry 1.0 began to be implemented into management programs in order to raise the efficacy of factories. This was done in order to improve the overall efficiency of the manufacturing sector. The division of labor, just-in-time production, and lean manufacturing concepts are just a few examples of production management strategies that have helped streamline the processes at their core, resulting in higher quality products and greater efficiency. The American mechanical engineer Fredrick Taylor is acknowledged as being a pioneer in the field of research into strategies for maximizing efficiency in the workplace. The third wave of the industrial revolution, also known as Industry 3.0, was ignited by advancements in electronic technology in the latter half of the twentieth century. As a result of the development and production of a wide variety of electronic devices, such as transistors and integrated circuits, which have significantly increased the level of automation in machines, there has been a significant reduction in the amount of labor required, an increase in speed and precision, and in some instances, the human agent has been completely replaced. In the 1960s, one of the very first electronic devices to announce the beginning of the era of industrial automation was the programmable logic controller, or PLC for short. The incorporation of these electronic devices into industrial systems resulted in the necessity of developing software systems to enable these devices, which in turn spurred the growth of the software development industry. Software solutions not only managed the hardware but also permitted a wide variety of management activities across the plant. These management operations included enterprise resource planning, inventory management, shipping logistics, product flow scheduling, and tracking. Further mechanization of the industry was accomplished with the use of electronic and computer technology. Since that time, the fields of electronics and information technology have seen significant advancements, and along with those advancements have come improvements in the software and automated processes that support these fields. The consistent pressure to reduce costs resulted in the relocation of many enterprises to nations with lower labor and material costs. The globalization of industry led to the development of a discipline known as supply chain management.

The 1990s were a time of phenomenal expansion for the internet and the telecommunications industry, both of which profoundly impacted how people communicate with one another and disseminate information. Other results include modifications in the paradigms used in manufacturing and the merging of physical and digital production processes that were previously treated as separate. The increase in the number of cyber-physical systems, also known as CPSs, has further muddied waters, which has triggered a wave of rapid technical innovation. Because of CPSs, machines can have more in-depth discussions with one another that are also more contextual, and they can do so over practically unlimited distances and without ever having to leave their own networks.

Industry 4.0 is a term used to describe the current trend of automation and data exchange in manufacturing technologies. It refers to the fourth industrial revolution, which builds upon the third industrial revolution (also known as the digital revolution) and the use of electronics, information technology (IT), and automated production.

Industry 4.0 is characterized by the integration of CPSs, the IoT, and cloud computing to create what is commonly referred to as a smart factory. These technologies allow for machines, systems, and products to communicate and exchange data with each other, leading to increased automation, efficiency, and productivity.

In Industry 4.0, machines are equipped with sensors and connected to the internet, allowing them to gather data about their own performance and the surrounding environment. This data can be analyzed in real-time, providing insights that can be used to optimize production processes, reduce waste, and improve quality control. Additionally, Industry 4.0 technologies can enable more flexible and customized production, as well as new business models such as product-as-a-service.

Overall, Industry 4.0 represents a significant shift in the way that manufacturing processes are designed and executed, with a focus on leveraging digital technologies to create more efficient, flexible, and sustainable operations. It makes use of CPSs in order to ease the sharing of data, the analysis of that data, and the guidance of intelligent actions across a wide range of industrial processes [1]. These clever devices are able to keep a watchful look out for potential issues and provide guidance on how to resolve them before the issues even arise. As a direct consequence of this, companies can be better prepared and have less downtime. The same dynamic approach could be beneficial in a number of different domains, including logistics, production scheduling, throughput time optimization, quality control, capacity utilization, and efficiency development, to name just a few. CPSs also allow remote visualization, monitoring, and management of an entire industry, which adds a whole new level to the manufacturing process. Management is simplified when everything, including equipment, people, procedures, and infrastructure, is integrated into a single connected system. As the technology-cost curve continues to steepen, rapid technological breakthroughs will emerge at progressively lower costs, fundamentally changing the industrial ecosystem in the process. The fourth iteration of the manufacturing sector, known as Industry 4.0, is still in its infancy, and the vast majority of companies are still in the throes of a transitional period as they integrate new technology. It will be necessary for industries to make the transition to the new methodologies as quickly as feasible in order to preserve their relevance and profitability. The fourth industrial revolution has arrived, and its influence will be felt for at least the next 10 years.

The fourth industrial revolution has the potential to usher in a number of remarkable technological advances in manufacturing settings. For instance, there are machines that are able to identify malfunctions and start up the necessary repair procedures on their own, and there are also self-organized logistics that can respond to unforeseen shifts in production. And it has the potential to completely transform the way that individuals do their jobs. Individuals can be enticed into more intelligent networks by Industry 4.0, which promises that their work will become more efficient as a result. The production environment has become more digitalized, which has allowed for more adaptable methods of communicating accurate information to the appropriate individuals at the appropriate times. Because more and more workers are using mobile devices both inside and outside of the factory, this suggests that technicians will have access to more up-to-date paperwork and service records for their equipment, just when and where they need it. This is because more and more workers are using mobile devices both inside and outside of the factory. Maintenance professionals would prefer solving problems rather than spending time investigating potential solutions to those problems. In a nutshell, the introduction of the fourth industrial revolution, has the potential to radically transform the appearance of industry as a whole. The process of digitizing production will have an effect on each stage of the supply chain, from the procurement of raw materials to the distribution of finished goods. Given these considerations, it is plausible to assert that it heralds the beginning of the fourth industrial revolution.

The adoption of CPS, such as the IoT and the Internet of Systems, is at the heart of the fourth industrial revolution. The IoT is a network of interconnected smart devices that enables each individual device to communicate with other devices on the network (i.e., send or receive data). Internet of Devices and Systems are owned by the businesses that are able to collect data from IoT networks in order to make independent decisions regarding the marketing campaigns, sales, and other aspects of the business. In the future, when the IoT is more widely adopted, intelligent devices will have greater access to data, which may enable them to make decisions and control critical business processes, such as supply chains, without any involvement from a human operator. Who we ask and what they think are the most important factors in determining whether or not autonomous machines are a good thing. Some people imagine a dystopian and hellish world straight out of a science fiction movie, in which robots have taken over all of the jobs, leaving humans jobless and miserable. Without work, our lives would lose all sense of purpose, which would increase the likelihood of drug addiction, violent acts, and widespread social unrest. Others, perhaps the majority, think that robots would free us from the tedious parts of our jobs, making us happier and more productive as a whole by allowing us to focus on the parts of our jobs that actually require human intelligence. When the fourth industrial revolution is fully developed, it will affect virtually every economic sector in the world.

To be more specific, the introduction of 4G mobile internet, artificial intelligence (AI), and automation, as well as big data analytics and cloud computing, are all major forces that are driving us toward the fourth industrial revolution. AI and automation are anticipated to have the most significant impact on employment rates around the globe, out of the four advancements listed above. The rise of AI and automation is expected to have an effect on one-fifth of the world's workforce, according to a recent study conducted by the McKinsey Global Institute [2]. This change is expected to have the greatest impact in industrialized nations such as the United Kingdom, Germany, and the United States. According to the World Economic Forum, more than 25% of companies anticipate that automation will lead to the creation of new roles. Furthermore, 38% percent of firms believe that AI and automation technology will enable employees to perform new professions that enhance productivity. Some people believe that the fourth industrial revolution will result in the creation of new job titles and functions, in addition to an increase in the use of freelancers and telecommuters in specialized fields. Therefore, with developments in technology and alterations in the requirements of businesses, employers may become more receptive to the concept of working from home and maintaining flexible work hours. When workers are given more leeway in terms of when, where, and how they finish their work, businesses can reap a number of benefits. It's possible that this will allow them to recruit employees from all over the world, increase the loyalty and dedication of the staff they already have, rapidly expand operations, and achieve levels of efficiency that were previously unimaginable. Employees will be happier and more committed to the company as a whole because they will have more time for personal pursuits and a healthier work-life balance as a result of not having to spend time commuting to and from work. It is evident that this phenomenon will have far-reaching repercussions given the prediction made by *The Economist* that during the fourth industrial revolution, half of all employment will be at risk of becoming automated. Certain sectors are more susceptible to being replaced by machines than others due to the fact that both humans and machines require specialized skill sets to do their jobs effectively. In light of the fact that an increasing number of manufacturing and agricultural jobs are already being eliminated as a direct result of increased automation, it is reasonable to anticipate that the number of full-time workers employed in these fields will decrease in the not-too-distant future. Robots were utilized in industries as early as the 1970s due to their capacity to do duties in a manner that was both more effective and secure. The Organization for Economic Cooperation and Development (OECD) recently came out with a list of jobs across a variety of industries that are at risk of being automated or eliminated entirely in the near future. The highest rankings go to jobs in the food service industry, construction, housekeeping, transportation, and farming.

In a nutshell, the IoT is a network that is interconnected and consists of physical things that are fitted with electronic sensors, actuators, and digital devices that are able to run specialized communication-enabling software. Every single one of them is connected to the rest of the globe by means of some kind of integrated networking, the internet being the one that is most often used. The first category includes things like coffee makers and light switches, while the second category includes things like pumps and motors. Devices like these are more widespread in the second category. This industry has been referred to as the IIoT in a number of the published works that have been done on the subject. Because it includes not only the internet but also embedded systems, the Internet of Manufacturing Services, the Internet of People, and the Internet of Things, it is also referred to as the IoE. Connectivity has been around for a while, but what makes Industry 4.0 revolutionary is that it integrates devices through the IoT, which opens the door to unprecedented levels of data collection and exchange. According to projections made by Gartner, a company that specializes in market research, there have been more than 11 billion connected devices in use all over the world by the end of 2018. This number is expected to have increased to approximately 21.8 billion by the year 2023. The internet has been an extremely important contributor to the growth and development of the industrial internet. At a more granular level, the management of the connections between a broad array of devices requires unique critical strategies that are necessary to do so in an environment that is always changing. In fact, prior to the widespread adoption of RFID tags, the term IoT referred to RFID-tagged linked devices that could be identified individually. This usage existed before RFID tags were widely used. According to recorded history, in the year 1982, students at Carnegie Mellon University connected a modified Coca-Cola machine to the World Wide Web. This way of identification functions as a tag, and instant tracking is made possible as a result of the distinctive nature of the method. After some time, this technique is developed further to include working together over a wireless network, Bluetooth, cellular networks, or near-field communication. According to van Kranenburg, the IoT is what he refers to as "a dynamic global network infrastructure with self-configuring capabilities based on standard and interoperable communication protocols where physical and virtual 'things' have identities, physical attributes, and virtual personalities, use intelligent interfaces, and are seamlessly integrated into the information network." Numerous IoT applications, including monitoring in fields as varied as manufacturing, the environment, transportation, and healthcare, are now in use. However, one should be aware that in order to connect billions of devices, there will need to be some standard protocols that all participants in the new era of Industry 4.0 would need to adhere to. Problems relating to privacy, security, and even the question of who actually owns what could emerge as a consequence of this. Therefore, cutting-edge research that draws from a number of different fields is constantly required if one wants to uncover unique solutions to important problems that exist today.

2.2   ANTENNA TECHNOLOGIES IN THE NEAR FUTURE

Traditional antenna architecture has exceeded its boundaries in various high-stakes industrial and aeronautical applications, such as 5G, SATCOM, the IoT, and radar [3]. One solution to this problem is to develop new types of antennas. However, a large number of companies are working on the development of novel techniques and materials that offer the ability to significantly improve antenna effectiveness and open up new application domains that were not viable in the past as a result of these limits.

Recent developments in the field of additive manufacturing, more commonly known as 3-D printing, have made it possible to materialize intricate RF designs. It has been demonstrated that a comprehensive characterization of the materials used in 3-D printing technologies is essential to the designing process, as well as the accurate forecasting of the performance of the resulting designs. Researchers have been able to learn more about the RF properties of the materials thanks to characterization, which, in turn, has encouraged the design of novel structures that would be impossible to build using production methods that are more traditionally used. Three-dimensional printing has made it possible to mass create traditional antenna forms using less material at a lower cost. This is made possible thanks to the advent of the technology. They are able to develop goods that are more iterative and intricate as a result of the adaptability of the 3-D printing technology. Utilizing this leeway to make more intricate RF parts, which can occasionally result in improved RF performance, is one technique to improve RF performance. Traditional machining techniques have their limitations when it comes to the production of RF goods, particularly when it comes to the creation of things with complicated shapes. To circumvent this limitation, complicated products are often built from a greater number of simpler subcomponents that are developed separately. This construction method is used in some cases. Because it does not have these limits, 3-D printing technique is able to produce finished devices using only a single part. This results in improvements in mass, cost, lead time, assembly quality, and RF performance. Another advantage of utilizing 3-D printing is a reduction in overall weight. Waveguide, filter, and antenna components of the C-band all the way up to the W-band have been used to successfully demonstrate the technology (4 to 110 GHz).

2.3   ANTENNAS SUITED FOR INDUSTRY 4.0

There are several types of antennas that are well-suited for use in Industry 4.0 applications [4], depending on the specific use case and requirements. Here are a few examples:

1. *Patch antennas*: These are small, low-profile antennas that can be used for wireless communication in confined spaces or where there is limited mounting space. They are often used in Wi-Fi, Bluetooth, or ZigBee networks, which are common in Industry 4.0 applications.

2. *Dipole antennas*: These antennas are commonly used in RFID systems for tracking and identification purposes. They can be designed for specific frequency bands, and can be used for short-range communication in indoor environments.

3. *Yagi antennas*: These are directional antennas that can be used for long-range communication in outdoor environments. They are commonly used in IoT applications for remote monitoring and control, such as in agriculture or environmental monitoring.

4. *MIMO antennas*: As mentioned earlier, MIMO antennas are used in wireless communication systems to improve the performance and reliability of data transmission. They are commonly used in Industry 4.0 applications that require high-speed and reliable wireless communication.

5. *LoRa antennas*: These antennas are designed for use with Lo-RaWAN networks, which are commonly used in IoT applications that require long-range, low-power communication. They are often used in smart cities, smart agriculture, and other applications that require long-range wireless communication.

Overall, the choice of antenna will depend on the specific requirements of the Industry 4.0 application, such as range, bandwidth, directionality, and power consumption.

2.4 METAMATERIAL-BASED ANTENNAS

In most cases, a metamaterial is produced by arranging a naturally occurring material in such a way that it is caused to display an electromagnetic response that is not ordinarily seen in the native state of the material. This enables the substance to be transformed into a metamaterial [5]. By creating periodic structures at scales lower than the wavelengths of the phenomena they impact, it is possible to fabricate materials with negative indices that govern electromagnetic energy in ways that are not possible with natural materials. These materials have the ability to govern electromagnetic energy in ways that are not possible with natural materials. Phase shifters found within the AESA's control circuitry allow for the array's orientation to be manually adjusted and, as a result, the beam's trajectory can be altered. AESAs that are based on metamaterials may steer the beam without the use of phase shifters, which eliminates a potential source of power loss, simplifies the process of waste heat dissipation, and reduces overall system complexity. Several different companies are currently employing the utilization of metamaterial structures that have been specifically built for this purpose.

After many years of experimentation with these structures, the company Kymeta came to the realization that metamaterials could be utilized to create holographic beams that could link to satellites and maintain the link while the antenna is in motion. This was the result of Kymeta's research into the potential of these metamaterials. The Kymeta mTenna® technology, in contrast to conventional antennas and phased array antennas, is constructed with an entirely unique manufacturing method and set of components. In mTenna technology, the metamaterial is actually a metasurface embedded within a glass framework. Their glass-on-glass construction is made on the same production lines as LCD flat screen televisions, which makes it suitable for low-cost, high-volume manufacturing. Additionally, the LCD flat screen televisions are manufactured on the same production lines. As a tunable dielectric, they make use of thin-film transistor liquid crystals. Kymeta uses a thin structure with elements made of tunable metamaterials to create a holographic beam that can transmit and receive satellite signals rather than reflecting microwaves like a traditional dish antenna or creating thousands of separate signals like a phased array. This allows the holographic beam to function similarly to a traditional dish antenna. Software is employed to guide the antenna, so there is no longer a requirement for mechanical gimbals to orient the antenna in the direction of a satellite. It is not necessary to utilize active phase shifters or amplifiers with this antenna. Transmit and receive data through a single aperture are two of the key features of this technique. Electronically controlled aiming and polarization, wide-angle scanning, and superb beam performance are some of its other features. First, electronically scanned antenna developed for mass production must feature extremely low–power consumption.

Another company, Echodyne, has also built metamaterial arrays for radar using antenna technology similar to that used by Kymeta. The company's primary focus is on radar. Echodyne's radar vision platform is a revolutionary sensor technology that, in addition to the radar's capabilities of all-weather, long-range, and ground-truth readings, is also capable of producing high-resolution photographs. The second half of radar vision is software that resembles computer vision in terms of classification, recognition, and perception. High-performance adaptive imaging radar hardware is only one-half of radar vision. Their electronically guided array radars based on metamaterials collect high-resolution data no matter the weather 24 hours a day, seven days a week, 365 days a year. This functionality is identical to that of conventional designs. The benefits that Kymeta's approach offers, such as high throughput production, commercial price, and a small and lightweight design, are also offered by these other methods. Their system is capable of switching in a timeframe of less than one millisecond, of steering in either direction, and of employing beam shaping and numerous beams in order to cover almost the entirety of both the northern and southern hemispheres. The operating frequency is 24 GHz, the range resolution is 3.2m, the azimuthal field of vision is 120 degrees, the elevation field of view is 80 degrees, the range is 3.4 km, and the velocity resolution is 0.9m per second. When it comes to the alternatives, the conventional radar in this region does not have the required resolution, while lidar and cameras have a limited range and are unreliable when the weather is bad. Echodyne's radar vision platform provides a new category of sensor technology that may be utilized by a wide variety of autonomous vehicles, ranging from drones to automobiles. The high-performance imaging radar may be utilized on commercial and compact platforms, which qualifies it for application in a wide variety of autonomous and unmanned vehicles and gadgets at a cost that is affordable.

2.5 TERAHERTZ ANTENNAS FOR IoT

Terahertz antennas are a type of antenna that operates in the terahertz frequency range, which spans from 0.1 to 10 THz. Terahertz antennas [6] have the potential to enable high-speed wireless communication for IoT applications, as they can provide significantly higher bandwidths than traditional RF antennas.

One of the main advantages of terahertz antennas is their ability to operate in the THz gap, which is the frequency range between the microwave and infrared bands. This frequency range has been largely unexplored until recently, but advances in terahertz technology have opened up new opportunities for high-speed wireless communication. Terahertz antennas can be used for a wide range of IoT applications, including high-speed data transfer, imaging, sensing, and spectroscopy. For example, terahertz antennas can be used for remote sensing applications in agriculture, where they can be used to detect soil moisture levels, crop health, and other environmental factors.

One of the challenges of terahertz antennas for IoT is their limited range, as terahertz signals are quickly absorbed by the atmosphere and other materials. Additionally, terahertz technology is still relatively new, and there are challenges with developing efficient and cost-effective terahertz components. Despite these challenges, terahertz antennas show great potential for enabling high-speed wireless communication in IoT applications, and there is ongoing research and development in this area to overcome the technical hurdles and bring terahertz technology into practical use. While terahertz technology is still relatively new, there are some practical applications of terahertz antennas in IoT that are currently being explored. Here are a few examples:

1. *Sensing and imaging*: Terahertz antennas can be used for sensing and imaging applications in IoT, such as detecting defects in materials or identifying hazardous substances. For example, terahertz imaging can be used to detect concealed objects or substances, such as drugs or explosives, which could be useful for security applications.

2. *Communication*: Terahertz antennas have the potential to enable high-speed wireless communication for IoT applications, as they can provide significantly higher bandwidths than traditional RF antennas. This could be useful for applications that require large amounts of data to be transferred quickly, such as video streaming or remote control of machinery.

3. *Medical applications*: Terahertz antennas can be used in medical applications for imaging and sensing, such as detecting skin cancer or monitoring blood sugar levels. For example, terahertz imaging can be used to identify the boundaries of tumors or to detect changes in tissue density, which could be useful for diagnosis and treatment planning.

4. *Environmental monitoring*: Terahertz antennas can be used for environmental monitoring applications in IoT, such as detecting gas leaks or monitoring air quality. For example, terahertz spectroscopy can be used to detect specific gases in the atmosphere, which could be useful for environmental monitoring and control.

Overall, while terahertz technology is still in the early stages of development, there are many potential applications of terahertz antennas in IoT that could enable new capabilities and improve existing applications.

2.6   FRACTAL ANTENNAS

An intricate pattern that is created by iterating a simple shape is called a fractal [7]. The application of fractal geometry results in the formation of an antenna that is referred to as a fractal element. Fractals, with their one-of-a-kind properties, can be put to use in the production of compact, high-performance antennas that can be as much as 75% more compact than their conventionally designed equivalents. There are a few basic benefits, the most popular of which are improved multiband performance, better gain, and broader bandwidth. Because the performance of fractal antennas is achieved through the geometry of the conductor, rather than through the accumulation of separate components or separate elements, which can increase complexity, potential failure points, and cost, fractal antennas may be more reliable than conventional antennas and may cost less than conventional antennas. Fractal antenna designs include dipoles, monopoles, patches, conformal antennas, biconical antennas, discone antennas, spiral antennas, and helical antennas. They were the first people to demonstrate wideband RF invisibility cloaking, and they did so by covering a sheet of mylar with a metal pattern in the shape of a fractal. They demonstrated that an attenuation of only a few decibels (dB) occurred behind the cloak for a signal with a bandwidth of 50% (from 750 to 1,250 MHz), whereas the attenuation would have been between 6 and 15 dB if the cloak had not been present. *Microwave Journal* lauded the innovative antenna-less technology developed by Fractus Antennas and published it in their October 2017 issue. This technology entails replacing a complex and typically one-of-a-kind antenna design with a commercially available, standardized, and small component known as an antenna booster. Because the antenna amplifier is a surface mount, chip-like device, it may be easily integrated onto a PCB by simply picking it up and setting it where it needs to go. It is designed to be used in applications related to mobile devices and the IoT, and it is formed from metalized ceramic layers with fractal forms that may be customized to meet a variety of design requirements. Not only is the concept of small chip antennas not new, but neither is the capability of the device to simultaneously function on a number of different frequency bands. When compared to the high-permittivity ceramics used in traditional miniature chip antennas, which enabled them to provide good performance for narrowband, single-frequency applications, the new boosters can provide full mobile performance across a wide range of frequency bands (for example, 698 to 2,690 MHz) using a single device. This is in contrast to the traditional miniature chip antennas, which were only able to provide good performance for narrowband, single-frequency applications. In order for the integration to be successful, the device in question needs to be able to function within the appropriate frequency bands; hence, a matching circuit is essential. The boosters may be manufactured in large quantities at a low cost since they are constructed using standard production procedures and components that are not particularly expensive.

2.7   PLASMA ANTENNAS

Plasma antennas (PSiAn) [8] offer a wide range of innovative plasma-silicon devices in order to supply the miniscule RF core that will be found in the future's smart antennas (PSiD). Due to the fact that the PSiDs are electronic, they are capable of performing beamforming and beam selection in a rapid manner. PSiDs are useful because they can perform the functions of RF switches, phase shifters, and attenuators, eliminating the need for many bulky and inefficient devices. PSiDs are able to be mass-produced with a high degree of accuracy and at a reasonable cost since they are manufactured using silicon integrated circuits. In contrast to RF MEMS, they can be switched while the device is still running and can withstand a significant amount of power. Beam steering in azimuth and elevation can be done with PSiAn by employing either a single PSiD or a cluster of PSiDs. Both approaches have their advantages. PSiDs are mounted on RF PCBs, and transmission lines are used to connect the device ports to conventional RF and antenna technologies such as LNAs, PAs, printed feeds, lenses, and reflectors. These steps are taken in order to produce efficient smart antennas with steerable narrow beams. PSiAn plasma antennas have a wide range of potential applications, including small cell backhaul in the V-band (60 GHz), gigabit wireless local area network (LAN, e.g., WiGig), intelligent transportation systems (ITS) in the 63-GHz range, and vehicle radar (77 GHz). They have also introduced their mmWave PSiAn, which is designed for usage in mobile devices such as smartphones and other consumer electronics. This technology makes use of directional beams to minimize interference, increase energy efficiency, and provides high throughput with low latency. The deployment of mmWave connection for smartphones and other mobile devices faces significant hurdles as a result of the ease with which fingers, hands, and bodies can block mmWave signals. In order to find a solution to this issue, it will be necessary to use a combination of distributed radiating elements and PSiDs. PSiDs perform the functions of a switch and beam former, allowing for the utilization of only those elements that are able to receive and transmit line of sight (LOS) or reflected signals. Plasma antennas have started modeling plasma-silicon corner antennas as drop-in replacements for array modules in the semiconductor sector recently. These antennas are intended to serve in this capacity. This method was a close depiction of the publicly known solutions from Qualcomm and Samsung, and it was one of the many conceivable modifications that may prohibit the antennas from functioning. These problems have been solved according to the array of plasma antennas that has been proposed, which also delivers the exceptional qualities of plasma silicon.

2.8 RECONFIGURABLE ANTENNAS

Reconfigurable antennas are antennas that can change their characteristics such as radiation pattern, frequency, and polarization in response to changing conditions or requirements. These antennas can be controlled through various mechanisms such as switches, varactors, or metamaterials.

There are different types of reconfigurable antennas based on the mechanism used to achieve reconfigurability [9]. Here are some of the commonly used mechanisms for reconfigurability:

1. *Switched beam antennas*: Switched beam antennas use multiple antenna elements that can be selectively switched on or off to steer the beam in a specific direction. By switching different combinations of antenna elements, the antenna can adapt to different directions or communication requirements.

2. *Frequency reconfigurable antennas*: Frequency reconfigurable antennas use mechanisms such as tunable capacitors, varactors, or phase shifters to adjust the resonant frequency

of the antenna. This allows the antenna to adapt to different frequency bands or communication standards.

3. *Polarization reconfigurable antennas*: Polarization reconfigurable antennas use mechanisms such as switches or rotatable elements to change the polarization of the antenna. This allows the antenna to adapt to different polarization requirements or to mitigate interference from other polarizations.

4. *Shape reconfigurable antennas*: Shape reconfigurable antennas use mechanisms such as shape memory alloys, electroactive polymers, or MEMS to change the physical shape of the antenna. This allows the antenna to adapt to different radiation patterns or to switch between multiple modes of operation.

Reconfigurable antennas have several advantages for IoT applications. They can adapt to changing communication needs, optimize the wireless link, and improve the energy efficiency of IoT devices. However, reconfigurable antennas also have some challenges such as increased complexity, cost, and power consumption. As such, the design and implementation of reconfigurable antennas require careful consideration of the trade-offs between performance, cost, and practicality for the specific application. Here are some examples of how reconfigurable antennas are currently being used in IoT applications:

1. *Smart home devices*: Reconfigurable antennas can be used in smart home devices, such as smart thermostats or security cameras, to adjust the radiation pattern and frequency based on the position and proximity of other devices in the home. This can improve the reliability and range of wireless communication between smart home devices.

2. *Connected vehicles*: Reconfigurable antennas can be used in connected vehicles to adjust the radiation pattern and frequency based on the location and orientation of the vehicle, as well as the presence of other vehicles and obstacles. This can improve the reliability and efficiency of wireless communication in vehicle-to-vehicle and vehicle-to-infrastructure applications.

3. *Wearable devices*: Reconfigurable antennas can be used in wearable IoT devices, such as fitness trackers or smartwatches, to adjust the radiation pattern and polarization based on the body orientation or proximity to other devices. This can improve the reliability and energy efficiency of wireless communication in wearable IoT devices.

4. *Agricultural sensors*: Reconfigurable antennas can be used in agricultural IoT sensors to adjust the radiation pattern and frequency based on the location and orientation of the sensor, as well as the presence of other sensors or environmental factors. This can improve the reliability and range of wireless communication in agricultural IoT applications.

5. *Smart city infrastructure*: Reconfigurable antennas can be used in smart city infrastructure, such as streetlights or traffic sensors, to adjust the radiation pattern and frequency based on the location and orientation of the device, as well as the presence of other devices or environmental factors. This can improve the reliability and efficiency of wireless communication in smart city applications.

Overall, reconfigurable antennas have the potential to improve the performance, flexibility, and reliability of a wide range of IoT applications by adapting to changing communication needs and environmental conditions. As such, there is ongoing research and development in this area to improve the performance and practicality of reconfigurable antennas for IoT applications.

2.9   GAP WAVEGUIDE

Gap waveguides [10] offer a variety of improvements over traditional transmission line and waveguide technologies, which make them excellent for the packaging of mmWave and terahertz circuits and components. These benefits can be broken down into categories. This technique makes use of a synthetic magnetic conductor to enable the contactless propagation of electromagnetic waves, which results in a significant reduction in the amount of transmission losses. The gap waveguide is made up of two surfaces, one of which is structured, and the other of which is flat. These two surfaces are in close prox-

imity to one another, but they are still separated by an air gap. The
pins that are distributed across the structured surface perform the
function of a shield, preventing electromagnetic waves from traveling
in unintended directions. This eliminates the requirement for perfect metallic contact, making it possible for the pins, rather than the
walls of a standard rectangular waveguide, to serve in the role of the
waveguide's walls. Because the waves are guided by ridges or grooves
within the pin structure, and because they travel through air, the pin
structure is able to achieve low power losses. Gap-based antennas
have losses that are roughly the same as those of rectangular waveguides, but they have losses that are more than 10 times lower than
those of microstrip lines, and more than three times lower than those
of substrate integrated waveguides (SIWs). Waveguide structures that
are multilayered and tightly spaced can be easily manufactured because metallic contact is not required for their construction. The layers of the antenna can be adhered to one another without the use
of any kind of fasteners, pressure, or heat. Plastic injection molding
combined with metallization or metal die casting are the two methods
that are utilized in the fabrication of the antenna components. These
methods allow for efficient mass production at a cheap cost. Because
of the low power losses, it is now possible to build broadband antenna arrays with strengths of up to 38 dBi and efficiencies of more
than 80%. A recently developed 38 dBi E-band antenna with ETSI
class 3 radiation pattern performance demonstrates the design versatility that comes with using multiple waveguide structures to permit
tweaking of the radiation pattern. The ETSI class 3 radiation pattern
performance is the highest level of performance that can be achieved
by an E-band antenna. Due to the fact that they possess several advantageous qualities, gap waveguides are an excellent choice for application in the production of active antenna systems. It is much easier
to combine gap waveguide-based antennas with PCB since there is
no need to establish an electrical connection between the PCB and
the antenna layers. In addition to this, the pin structure protects the
active components from electromagnetic interference and stops any
signals from propagating through the PCB's substrate. By doing away
with the need for through holes and shielding walls, additional space
can be made available on the circuit board for the components and
connections that are really being used. Die-cast antenna layers provide effective cooling from both directions to the active circuits, al-

lowing for optimal performance. It is helpful to do this when merging high-power amplifiers and complementary metal-oxide semiconductor (CMOS)-based control circuits onto a single circuit board because these two types of circuits have different cooling requirements.

2.10 NEW ANTENNAS WITH MIMO CAPABILITIES FOR IoT APPLICATIONS

MIMO, as defined in [11], is a technology used in wireless communication systems to improve the performance and reliability of data transmission. A MIMO antenna system uses multiple antennas at both the transmitter and receiver to increase the capacity and quality of the wireless connection.

In IoT applications, MIMO antennas can be used to enhance the communication between IoT devices and the network, especially in environments where the wireless signals are weak or have high interference. MIMO antennas can improve the reliability and speed of wireless communication between IoT devices, leading to more efficient and effective IoT applications.

MIMO antenna systems work by exploiting the spatial diversity of wireless signals. By using multiple antennas, the transmitter can send multiple signals simultaneously, which can then be received by multiple antennas at the receiver. This allows for the separation and decoding of the different signals, leading to a more reliable and faster data transmission.

In summary, MIMO antennas are used in IoT [12, 13] to improve wireless communication performance by increasing the capacity and quality of the wireless connection, especially in challenging environments where the signals are weak or have high interference.

The following are some examples of antennas with MIMO capabilities that are suitable for IoT applications:

1. *Poynting XPOL-1-5G*: This is a high-gain, dual-polarized antenna designed for 5G applications, but it is also suitable for other IoT applications. It supports both MIMO and LTE Advanced Pro, and it provides excellent signal quality and coverage. It is designed for outdoor use and can be mounted on a pole or wall.

2. *Laird Connectivity S24517PTN83N300*: This is a low-profile, omnidirectional antenna that supports MIMO and is ideal for IoT applications that require a compact design. It operates on the 2.4-GHz and 5-GHz bands, and it provides a gain of up to 3 dBi. It can be mounted directly on the device or on a small ground plane.

3. *Taoglas Apex 2J*: This is a high-performance, dual-band antenna that supports MIMO and is designed for IoT applications that require a compact, low-profile design. It provides a gain of up to 4 dBi and operates on the 2.4-GHz and 5-GHz bands. It is suitable for use in indoor or outdoor environments.

4. *Antenova M20047-1*: This is a compact, low-profile antenna that supports MIMO and is designed for IoT applications that require a small form factor. It operates on the 2.4-GHz and 5-GHz bands, and it provides a gain of up to 4.5 dBi. It is suitable for use in a variety of IoT applications, including smart homes, wearables, and industrial monitoring systems.

These are just a few examples of antennas with MIMO capabilities that are suitable for IoT applications. There are many other options available, so it's important to consider your specific application requirements when selecting an antenna.

2.11 CONCLUSION

There are a number of exciting new technologies that have the potential to completely revolutionize antenna design in the not-too-distant future. Some examples of these technologies are 3-D printing, metamaterials, and fractal antennas. They will make it possible to create unique antenna designs and applications, both of which were previously unachievable with the technology that was available. The unconventional approaches that have been adopted will result in the resolution of a great deal of the problems that have plagued recent implementations of 5G, IoT, SATCOM, and radar.

References

[1] Alexiou, A., and M. Haardt, "Smart Antenna Technologies for Future Wireless Systems: Trends and Challenges," *IEEE Communications Magazine,* Vol. 42, No. 9, Sept. 2004, pp. 90–97, doi: 10.1109/MCOM.2004.1336725.

[2] Bughin, J., E. Hazan, S. Lund, P. Dahlström, A. Wiesinger, and A. Subramaniam, "Skill Shift: Automation and the Future of the Workforce," McKinsey Global Institute: McKinsey & Company, 2018.

[3] Jokanovic, B., V. Milosevic, M. Radovanovic, and N. Boskovic, "Advanced Antennas for Next Generation Wireless Access," *2017 13th International Conference on Advanced Technologies, Systems and Services in Telecommunications (TELSIKS),* Nis, Serbia, 2017, pp. 87–94, doi: 10.1109/ TELSKS.2017.8246235.

[4] Parthiban, P., "IoT Antennas for Industry 4.0–Design and Manufacturing with an Example," *2020 IEEE International IoT, Electronics and Mechatronics Conference (IEMTRONICS),* Vancouver, B.C., Canada, 2020, pp. 1–5, doi: 10.1109/IEMTRONICS51293.2020.9216349.

[5] Dong, Y., and T. Itoh, "Metamaterial-Based Antennas," in *Proceedings of the IEEE,* Vol. 100, No. 7, July 2012, pp. 2271–2285, doi: 10.1109/ JPROC.2012.2187631.

[6] Teng, F., J. Wan, and J. Liu, "Review of Terahertz Antenna Technology for Science Missions in Space," *IEEE Aerospace and Electronic Systems Magazine,* Vol. 38, No. 2, Feb. 1, 2023, pp. 16–32, doi: 10.1109/MAES.2022.3222291.

[7] Werner, D. H., and S. Ganguly, "An Overview of Fractal Antenna Engineering Research," *IEEE Antennas and Propagation Magazine,* Vol. 45, No. 1, Feb. 2003, pp. 38–57, doi: 10.1109/MAP.2003.1189650.

[8] Anderson, T., *Plasma Antennas,* Norwood, MA: Artech House, 2011.

[9] Haupt, R. L., and M. Lanagan, "Reconfigurable Antennas," *IEEE Antennas and Propagation Magazine,* Vol. 55, No. 1, Feb. 2013, pp. 49–61, doi: 10.1109/MAP.2013.6474484.

[10] Shi, Y., et al., "Gap Waveguide Technology: An Overview of Millimeter-Wave Circuits Based on Gap Waveguide Technology Using Different Fabrication Technologies," *IEEE Microwave Magazine,* Vol. 24, No. 1, Jan. 2023, pp. 62–73, doi: 10.1109/MMM.2022.3211595.

[11] Prakash, P., G. Manoj, and J. Samson Immanuel, "MIMO Antenna System for IoT Applications (5G)," *2022 6th International Conference on Devices, Circuits and Systems (ICDCS),* Coimbatore, India, 2022, pp. 178–182, doi: 10.1109/ICDCS54290.2022.9780845.

[12] Adcock, M. D., R. E. Hiatt, and K. M. Siegel, "The Future of Antennas," *Proceedings of the IRE,* Vol. 50, No. 5, May 1962, pp. 712–716, doi: 10.1109/ JRPROC.1962.288104.

[13] Kulkarni, P., and R. Srinivasan, "Compact Polarization Diversity Patch Antenna in L And WiMAX Bands for IoT Applications" *AEU–International Journal of Electronics and Communications,* Vol. 136, 2021, 153772, ISSN 1434-8411, https://doi.org/10.1016/j.aeue.2021.153772.

3

RF AND MICROWAVE ASPECTS OF IoT

3.1 INTRODUCTION

The IoT transforms everyday objects into intelligent systems that can transmit and analyze environmental data. The IoT, to put it simply, is a complex system made up of real-world items that include a variety of embedded sensors that are carefully created and set up with a secure network. Depending on the type of task the device performs, the data gathered from each smart device is processed in a centralized system. A connected smart house with all electronic components able to talk with one another and transfer information as needed and would be provided via IoT. For instance, sensors in the wall will be able to recognize one's presence in the space and regulate the temperature of the air conditioner. When someone approaches the door, smart sensors linked to it will assist in opening and closing the door. It may also set off other actions, such as turning on or off the room's lights or adjusting the airflow. Smart water level sensors can effectively monitor water use and regulate water flow. The washing machines are equipped with smart devices that can detect the detergent level and sound an alarm. It will be able to let the service staff know when something needs fixing during a breakdown.

3.2 IMPORTANCE OF RF AND MICROWAVE TECHNOLOGIES IN IoT

RF and microwave technologies [1] are essential for enabling wireless communication in the IoT ecosystem. IoT involves the interconnection of a vast number of devices, sensors, and systems, which requires reliable and efficient wireless communication to operate effectively. RF and microwave technologies [2] offer several advantages for IoT applications, including:

1. *Wireless connectivity*: RF and microwave technologies enable wireless communication between IoT devices and systems, eliminating the need for physical connections, such as cables or wires. This feature allows for greater mobility and flexibility in deploying IoT devices and systems.

2. *Low power consumption*: RF and microwave technologies are designed to operate on low power, making them ideal for IoT devices that have limited power sources, such as batteries or energy-harvesting systems.

3. *High data rates*: RF and microwave technologies offer high data rates, enabling the transmission of large amounts of data quickly and efficiently. This feature is essential for IoT applications, where data is constantly being generated and transmitted.

4. *Long-range communication*: RF and microwave technologies can transmit data over long distances, making them ideal for IoT applications that require remote monitoring or control.

5. *Compatibility with existing infrastructure*: RF and microwave technologies are compatible with existing wireless infrastructure, such as Wi-Fi and cellular networks, making it easier to integrate IoT devices and systems into existing networks.

Overall, RF and microwave technologies play a critical role in enabling wireless communication in IoT applications, making them essential for the development and deployment of IoT systems and devices.

3.3 SCOPE OF RF TECHNOLOGY IN IoT

The majority of IoT applications fit well into RF technology, which is frequently one of the key elements. RF devices are used by wearable technology to connect to smartphones and other smart devices. Real-time monitoring systems for medical equipment frequently use wireless technologies. RF transmitters and receivers are used in smart home technology to establish Wi-Fi or Bluetooth connections with the internet, as seen in Figure 3.1. Newer generation household appliances already include wireless connectivity.

In industries, supply chain management, smart farming, and many other uses, RFID sensors are unavoidable. In remote sensing, information is transmitted wirelessly to a different site for analysis. Temperature, humidity, and fire sensors are built into wireless transceivers and are essential parts of industrial IoT applications.

3.3.1 Wearable Technology

Young people and the elderly have already become very accustomed to wearing technology as seen in Figure 3.2. These smart devices are a must-have due to their convenience. Smartwatches, for example, are capable of much more than merely keeping time. A good smartwatch can track your heartbeat, blood pressure, sleeping patterns, and other

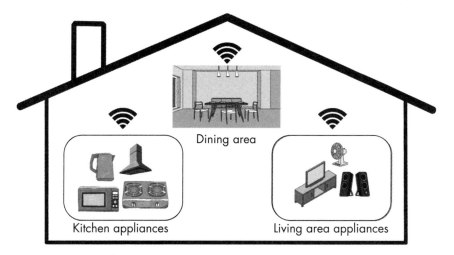

Figure 3.1 Smart home communication.

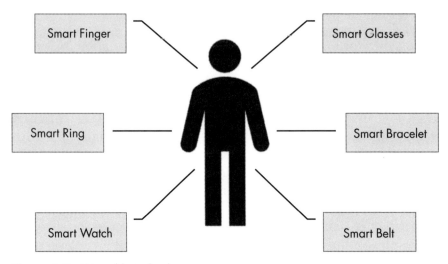

Figure 3.2 Wearable technology.

health indicators. In the not-too-distant future, a doctor will be able to evaluate patient data using wearable smart gadgets that have been specifically created for that purpose. Even in the case of life-threatening medical situations like cardiac arrest or stroke, it might send messages requesting assistance from medical services.

Most importantly, by tracking everyday activities, these smart devices can aid in enhancing a person's state of health. Virtual reality technology and devices like Google Glass have already captured our hearts.

3.3.2   Smart Cities

With cutting-edge wireless technology, smart cities will be the next big thing this decade, creating millions of opportunities for investors and ultimately for consumers. The concept is better depicted in Figure 3.3.

The way we manage our security, trash management, and pollution control systems will all be completely transformed by IoT. Better social security and disaster management are made possible by real-time monitoring systems.

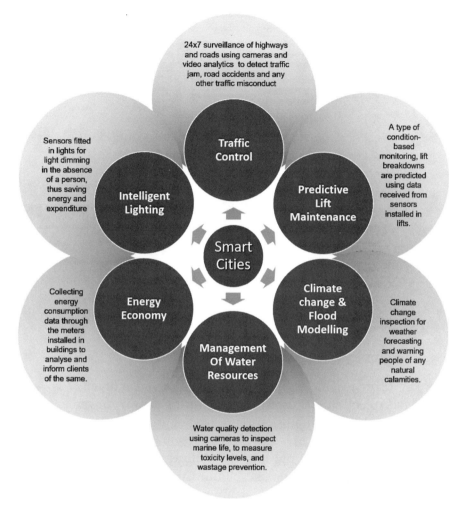

Figure 3.3 Smart cities.

3.3.3 Industrial and Automobile

IoT in business will present numerous opportunities and options to be integrated to new goods. The idea of a smart car was novel at the start of this century. Now that our cars may be connected to smartphones, we can utilize specialized software to carry out specific functions. These facilities are seen in Figure 3.4.

Numerous in-car sensors gather data and transmit it to a service team or manufacturer's database. The manufacturers will be able to

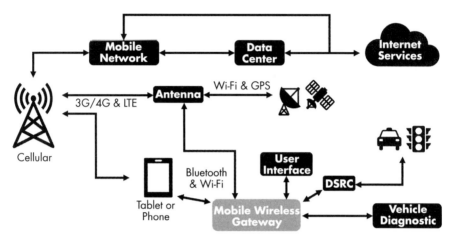

Figure 3.4 Smart car.

track and keep track of how each item performs with the use of this data. The design team will be able to keep refining their product with its assistance.

3.3.4 Supply Chain and Retail Management

The key to increasing supply chain management efficiency is real-time data collection. IoT assists in monitoring supply movement, and analytics of the same helps increase output and productivity. Effective supply chain and inventory management, as seen in Figure 3.5, is made possible by well-optimized mobile applications coupled with a central control system on the cloud. Customers and service providers can easily access it at their fingertips. Retail has undergone a change due to innovative price tags. Tags are always changing in response to market demands and trends.

Big businesses employ data analysis to optimize and forecast market trends that help them further develop their systems and satisfy client needs.

Figure 3.5 Supply chain management.

3.3.5   Healthcare

One of the most crucial fields for technological advancement is healthcare. A multibillion-dollar sector, healthcare needs higher quality and ease for all of its users. IoT will significantly contribute to raising the standard of healthcare services as depicted in Figure 3.6.

Doctors and other medical professionals can remotely monitor a patient's condition in real time. Each participant is integrated utilizing a mobile device. With the use of video conferencing, medical professionals can communicate with one another and, even from a distance, witness operations. Research and medical corporations can gain insights and enhance the quality of drugs, devices, and services by collecting and analyzing data.

3.3.6   Energy Management

A sophisticated energy management system that is integrated with appliances automatically detects what needs to be done and notifies users. Customers will find it simple to use their smartphone to schedule and control their energy-efficient home appliances and devices.

All home appliances, including air conditioners, refrigerators, and washing machines, eventually are able to interact and warn

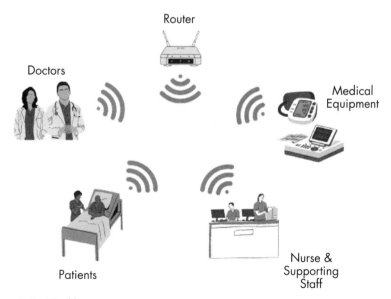

Figure 3.6   Healthcare services.

people in particular situations. Utilizing smart sensor technologies, IoT assists enterprises in managing energy efficiently, as seen in Figure 3.7. It facilitates better energy distribution and additional cost reduction. Power grids employ smart sensors to efficiently monitor and analyze data.

3.3.7 Smart Farming and Agriculture

Through the use of remote sensing technologies, massive amounts of data from a wide geographic area can be collected quickly and transmitted for additional processing. Farmers will be able to identify crop anomalies and significant damages that could affect a large area, and

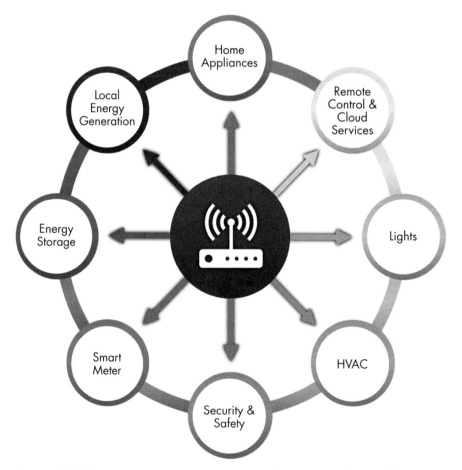

Figure 3.7 Energy management systems operated using a wireless module.

take appropriate action to move on to the next level. Farmers will be able to execute controlled irrigation and water management in accordance with various weather conditions thanks to IoT technology as in Figure 3.8. Temperature and moisture sensors monitor the environment in real time and send alerts when more action is required.

Smart sensor technology, which gathers data and transfers it to processing stations, enables IoT. A smart device with communication and sensing capabilities will improve the customer experience. IoT device security and safety concerns continue to be a topic of conversation. IoT creates chances for engineers, new tech businesses, and investors to study, examine, and upgrade legacy systems in order to make them smarter, more secure, and more energy efficient. We can quickly link our smart IoT devices to one another using RF technology without the need for laborious configuration.

3.4   RF FUNDAMENTALS FOR IoT

The IoT and the emergence of the IIoT are causing a wave of innovation across the industry, promising to improve asset utilization, improve

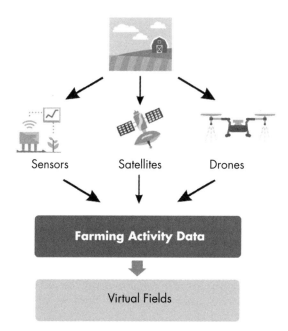

Figure 3.8   Smart farming.

process efficiency, and boost productivity by connecting an ecosystem of sensors, devices, and equipment to a network. Over 35 billion connected "things" are expected by 2025, with the potential to change how business is performed. Sensors, gateways, infrastructure, and, of course, the IoT ecosystem's most eye-catching component, big data and analytics, comprise the IoT ecosystem.

Because of the modeling of data enabling predictive analytics, organizations may swiftly diagnose and fix not just their sensor networks from a predictive maintenance standpoint, but also their operations, eliminating needless usage of energy and/or raw materials.

Although the ecosystem can be hardwired, a hybrid approach is frequently employed [3]. High-speed, broadband lines, wireless sensor networks (WSN), and the majority of network designs are included. Despite the fact that RF technology is a part of our daily lives and has been employed in the IIoT for decades in some of the harshest conditions, many people still regard wireless technology as magical. Though he was a gifted writer and visionary, Arthur C. Clarke is most recognized for his law that "any sufficiently advanced technology is indistinguishable from magic" from his book *Profiles of the Future.* Wireless (or RF) communications unquestionably fall under that definition.

Basis: To put it simply, wireless transmissions are RF signals that move according to attributes known as propagation characteristics. An antenna on the transmitter end emits RF that is picked up by an antenna on the receiving end. The true data is modulated on one end, then transferred over the air and demodulated on the other. The RF signal's propagation qualities are determined by frequency, wavelength, and amplitude. Depending on the technology and region of the world where the system is installed, regulatory organizations for that nation will limit the frequencies that can be used. In the United States, for example, the FCC designated the ISM band, permitting license-free operation in the 900-MHz, 2.4-GHz, and 5.8-GHz bands. Other countries, on the other hand, may use 900 MHz for their domestic cellular band or even free-to-air television. There are also licensed bands above 5.8 GHz and below 900 MHz. The propagation characteristics stated previously apply regardless of the band chosen. The standard operating range in the 900-MHz band, which is one of the most popular IoT bands, is 902 to 928 MHz. Hertz (Hz) represents frequency or cycles per second, while the "M" in front of it denotes a factor of one million. As a result, 900 MHz is equivalent to 900 million cycles per second. Wi-Fi's most common frequencies are 2.4 GHz and 5.8 GHz, where the "G" prefix stands for giga, or a factor of one billion. A wavelength is the distance between the beginnings of two cycles and is typically represented by the Greek letter lambda (λ). The abovementioned frequency is inversely proportional to wavelength. The wavelength becomes shorter as the frequency increases. On the other side, as frequency decreases, the wavelength lengthens. When the frequency is known, the wavelength can be estimated by dividing the speed of light by the frequency, which is rounded up to 300 billion meters per second. At 900 MHz, the wavelength is 33.3 cm, or about 13 inches. Understanding wavelength is critical for designing antennas, which will be discussed later.

The final RF propagation characteristic [4] is amplitude. Amplitude is the height or peak of the wave and is a function of power. With increased output power or amplitude, the peak rises. Almost every IoT technology offers completely adjustable output power. This is essential because, as previously stated in respect to Arthur C. Clarke's law, many people believe that the greater the output power, the better. Higher output power can sometimes cause self-interference and raise the total noise floor. In other cases, a higher output power is just required to drown out background noise. RF power is measured in terms of output power with two units on two different scales. The first use of a linear scale is based on milliwatts (mW). On a linear scale, the reference point is zero, but it masks any gain or loss relative to the total. The second scale is relative and in decibels (dB), with the measurement serving as the reference point. A decibel is simply a logarithmic scale ratio that can be misleading because it is purely relative and has no clear baseline against which to assess gains or losses. We simplify our calculations by measuring power in decibels, which allows us to add and subtract our gains and losses. The most common instances of referenced dB values are dBm and dBi. While dBi measures antenna gain in comparison to a fictitious, ideal isotropic radiator, dBm measures absolute power in relation to 1 mW. The rule of 3s and 10s is arguably the most important thing to remember right now in the world of RF.

The following are the five basic steps to perform any RF math calculation:

0 dBm = 1 mW (starting point);

30 dBm = 1W increase by 3 dBm, power in mW doubles;

Decrease by 3 dBm, power in mW is halved;

Increase by 10 dBm, power in mW is multiplied by 10;

Decrease by 10 dBm, power in mW is divided by 10.

3.4.1   Sensitivity

When evaluating wireless technology, the sensitivity specification is equally as important as the actual output power. We are all aware that it is more important to hear and, more importantly, understand what others are saying in daily interactions than it is to speak as loudly as possible. Sensitivity provides this hearing requirement. Receiver sensitivity is the lowest power level at which a receiver can identify an RF signal and, analogous to being able to follow a conversation, demodulate the data. A bit error rate (BER) parameter is normally placed after the receiver sensitivity definition. It is common to use the expression 10 to a negative power to describe the ability to understand what the other device is broadcasting. For instance, a BER of 10-4 means that, for every 10,000 bits transmitted, there is a 10% chance that a mistake would be made. Similar to output power, the 3s and 10s rule also holds true in this situation. The difference between a receiver with a sensitivity level of –110 dBm and one with a standard of –107 dBm is initially seen as being only 3 dB. The receiver with a –110 dBm specification can pick up signals that are only half as strong as those picked up by a receiver with a –107 dBm standard. Since receive sensitivity indicates how weak a signal may be effectively received, the lower the power level, the better. This suggests that as the number's negative absolute value increases, so does the receive sensitivity. 0 dBm = 1 mW (starting point) power in mW doubles at 30 dBm = 1W, increasing by 3 dBm. The power is halved in mW and lowered by 3 dBm. The power is raised by 10 mW, and the corresponding increase in dBm is 10 dB. To lower power by 10 dBm, multiply the power in mW by 10.

3.4.2   Use in Real Life

Regardless of the wireless technology employed, link budgets are essential in determining if a system will work. The link budget in a communications system accounts for all gains and losses from the transmitter to the receiver through the medium (free space, cable, waveguide, fiber, etc.). It accounts for antenna gains, additional losses, and the broadcast signal's attenuation brought on by propagation. The following is a simple link budget formula: Gains (dB) = Received Power (dB) + Transmitted Power (dB) + Losses (dB).

3.4.3   Transmitter Output Power

The transmitter output power will vary based on the technology being utilized and the country it is being used in. With intrinsically safe, battery-powered WSN, reducing energy consumption is crucial to extending battery life. Output power may only be 10 dBm as a result (or 10 mW). However, other systems, such as licensed radio systems, have the potential for significantly higher output power. Hypothetical example: 30 dBm (1W).

3.4.4   Transmitter Loss

The system will undergo losses from the coaxial cable as well as insertion losses due to jumpers, adapters/connectors, lightning arrestors, and so on from the RF connector of the transmitter to the antenna. It is crucial to review the coaxial cable manufacturer's specifications before remotely placing an antenna. The system may suffer a loss if the wrong cable is used, which will keep it from functioning properly. For instance, utilizing the same LMR400 low loss cable at 900 MHz results in a loss of 3.9 dB/100 ft while doing so at 2.4 GHz causes a loss of 6.8 dB/100 ft[1]. Example: 3.9 dB from a 10' LMR400 and two connectors at .2 dB each results in a loss of 4.3 dB.

---

1. http://rfelektronik.se/manuals/Datasheets/Coaxial_Cable_Attenuation_Chart.pdf.

3.4.5 Antenna Gain

Antennas are a separate topic all by themselves and are arguably one of the most challenging components of an RF system. Numerous books that go into great detail about antenna theory and design have been written. But an antenna is a device that transforms electrical currents into radio waves at the transmitting end and back again at the receiving end. Omnidirectional and directional antennas are the two most popular types of antennas. A directional antenna will focus its energy in one direction, as opposed to omnidirectional antennas, which emit energy in a 360°-pattern. In point-to-multipoint systems, the gateway (or access point) and repeater sites often employ omnidirectional antennas, whereas the remote sites typically utilize directional antennas, particularly in those applications where remote transmitters are located a great distance from the gateway. Every antenna has a gain specification that is commonly represented in dBi and, as was already established, refers to an idealized isotropic antenna. As gain rises, the directionality also does in directional antennas.

Effective isotropic radiated power, often known as EIRP, is the amount of energy that the antenna radiates, and is responsible for all gains and losses at that end of the system. The transmitter in our fictitious example up to this point has an output power of 30 dBm, cable and connector losses are 4.3 dB, and we'll be using an antenna with a gain of 3 dBi as shown in Section 3.4.6. The EIRP is therefore 28.7 dBm.

3.4.6 Hypothetical Example

Gain of 3-dB loss in free space path: RF attenuates over a distance similarly to how our voices do (the farther away you are from someone, the harder it is to hear them). In essence, the waveform's geometric spreading over distance reduces the signal's strength.

According to physics, the frequency attenuates more quickly the higher it is. As a general rule, range is halved when frequency is doubled. Numerous online calculators are available to determine the loss based on frequency and distance, but the following formula derived from the Friis transmission formula can be used to determine free space path loss (FSPL): FSPL = 36.56 + 20Log10 (frequency in MHz) + 20Log10 (distance in miles). It is crucial to remember that energy emitted from an antenna is not a laser beam traveling straight down. The Fresnel zone is where it radiates in a parabolic or elliptical pattern. The phrase "with clear line of sight" is included in almost all RF range requirements, which suggests that the Fresnel zone must be taken into account. The number of Fresnel zones is theoretically endless and each one has a smaller effect on the link. In Fresnel zone 1 (F1), which is most affected by obstructions, most path study or propagation software defaults to 60% clearance. Similar to holding a notebook in front of your lips as you speak as opposed to holding it at arm's length, the wireless link is more affected by an obstacle the closer it is to the radiating antenna. *Hypothetical illustration receiver antenna gain*: The same rules apply. 110 dB of loss (5 kilometers at 900 MHz). *Imaginary example*: 3 dBi of gain receiver loss: The same guidelines apply. A hypothetical loss of 4.3 dB. After all gains and losses, the received signal strength at the receiver is known as the received signal strength (FSPL of −110 dBm plus 27.4 dB of system gain). *Hypothetical example*: −82.6 dBm (the expected received strength level). Receiver sensitivity is the threshold or lowest power level at which a receiver can detect an RF signal, as was previously discussed. A reliable link cannot be created if the received signal strength is equal to or lower than the receiver. Contingency or margin must be taken into account as a result.

Hypothetical example: −110 dBm (receiver sensitivity, specified from manufacturer). Last but not least, the abovementioned contingency is the fade margin, which is the margin (given in dB) between the received signal and the receiver sensitivity. This buffer can aid in overcoming problems posed by the terrain as well as attenuation caused by obstructions, rain, and climatic conditions. There are varying views on the ideal fade margins, like there are for the majority of RF characteristics. A durable system with a high level of reliability can be created by designing systems with 20 dB of margin.

It is crucial to remember that receiver sensitivity does not, by itself, indicate the weakest signal that can be safely decoded. We have to deal with noise in the actual world. The system may be constrained by the noise floor rather than the receiver sensitivity as a result of highly saturated locations. A spectrum analyzer takes a screen capture of the ambient noise floor.

In this particular area, the calculated performance did not match the actual performance. As can be seen, the background noise level for the 900-MHz frequency was roughly −95 dBm. The noise floor therefore has a huge impact on our overall performance when applied to our hypothetical case, and instead of having a highly envious margin of 27.4 dB, it becomes 12.4 dB. Increasing the receiving antenna's gain won't help in severe circumstances of a high noise floor. The only way to get rid of the noise is to boost the transmitter's EIRP (within legal limitations). If this problem were to arise, spectral analysis in other bands might produce workable answers in those bands.

3.4.7 Conclusion

The IoT taps on the potential of interconnected systems to forecast, learn from, and make business choices in real time. However, you must have connected devices before you can have connected systems. IoT is not at all new, despite the fact that it may look like a buzzword or notion that is recent. For many years, businesses have used technology to track, evaluate, and manage their most important assets. The fast growth of invention, however, is novel. IoT carries with it an enormous rise in demand for wireless technology due to the acceleration of innovation. Understanding RF basics will guarantee proper application and execution of wireless systems, from sensor to gateway data communications, to IoT architecture. There are various industrial and commercial IoT applications, and dependable IoT links must be created through successful wireless data link design. Gains and losses are what matter most in the end. This high-level summary of the principles should help remove some of the subject's mysticism.

3.5   WHAT IS RF IN IoT?

The IoT is based on RF technologies [5, 6] that enable the IoT's long-range, low-power capabilities. IoT is evolving for the better with the arrival of 5G RF technology, but for some, utilizing the most recent and finest features may be challenging. First off, it's important to note that many different technologies are based on RF. RF technologies in the IoT refer especially to the usage of RF signals that allow for wireless data exchange between IoT devices! In actuality, RF is what makes the IoT possible. Therefore, choosing the right RF communication type for IoT is crucial! The RF communication link is as seen in Figure 3.9.

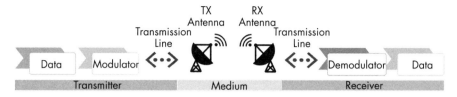

Figure 3.9 RF communication.

RF signals are generated by an antenna on a transmitting device and are received by an antenna on a receiving device. The signal is subsequently decoded into the payload data at the other end using demodulation. RF communication is frequently exemplified by Bluetooth, Wi-Fi, and LR/LoRaWAN.

Furthermore, there are a lot of reasons why RF in IoT is popular and acceptable, including:

1. Little power usage;
2. Wide operating bandwidth;
3. Fast data transfer rate;
4. Does not require LOS and penetrates walls.

3.6 TRANSFORMING IoT WITH 5G RF

The 5th generation mobile network, or 5G, is the replacement for the 1G, 2G, 3G, and 4G radio communication networks as depicted in Figure 3.10.

Will 5G RF technologies improve the functionality of IoT systems? Yes, in a lot of ways! Faster, more dependable, and secure wireless transmission made possible by 5G RF let us make great strides with IoT gadgets like smart cars, energy grids, and even AI-enabled machines and robots! Future IoT networks supporting billions of devices will be made possible by 5G RF on a scale never previously possible.

When contrasting 4G with 5G RF technologies, there may be some significant areas of uncertainty. It's crucial to remember that 5G is not fundamentally different from 4G despite requiring new technology and infrastructure.

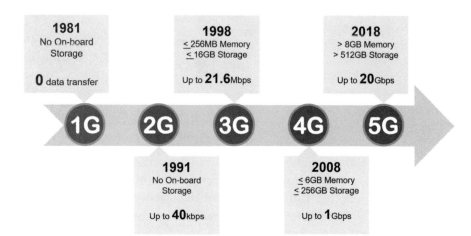

Figure 3.10 Evolution of communication networks.

5G NR, also known as new radio [7], is the new international standard for a unified wireless air interface and is the foundation of 5G. Through innovative methods like OFDM to prevent interference, it also considerably enhances 4G capabilities. In addition, 5G makes greater use of spectrum resources than 4G did, using a wider range of RFs to achieve flexibility across a wider range of applications. These applications can be summarized in three major groups as seen in Figure 3.11.

3.6.1 Enhanced Mobile Broadband

The primary goal of 5G's enhanced mobile broadband is to increase data capacity in high bandwidth applications up to 20 Gbps. Applications like consumer video streaming and security video surveillance will profit in this situation. Fixed wireless mmWave, beam steering, and other 5G technologies fall under this category. LTE-A and other 4.5G standards can also be included.

3.6.2   Low Latency

Low-latency applications are those that have strong requirements for quick reaction times rather than a significant amount of bandwidth.

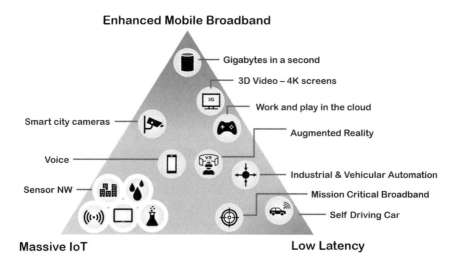

Figure 3.11   5G applications.

Through massive MIMO, 5G has the ability to 10× outperform 4G standards by reducing latencies to as little as 1 ms (MIMO). The possibilities are boundless, but mission-critical IoT devices like industrial safety mechanisms or cutting-edge healthcare systems stand to profit.

3.6.3   Massive IoT

Massive IoT is distinguished by two features: great range and low power. In addition to enabling scaling down of data speeds and power needs for usage in lean, effective IoT communications, 5G was created to enable seamless connectivity in devices of all sizes. CAT.M and NB-IoT are two examples of 5G standards that have been created to effectively coexist with 5G NR in order to meet cellular IoT requirements for the foreseeable future.

LR and LoRaWAN are excellent solutions for massive IoT applications even though they are not strictly categorized as 5G technology. The most cost-effective method for handling large-scale IoT that doesn't require high bandwidth, LR is nevertheless capable of handling thousands of connections at vast distances of up to 15 kilometers.

3.7   5G RF USE CASES IN IoT

Better 5G connectivity will only serve as a catalyst to propel the IoT to the next level since it is already evolving at an astonishing rate. Here are a few application scenarios for the IoT that will profit from 5G RF technologies [8].

3.7.1   Remote Robotics

Remote robotics is no longer so far off from reality thanks to 5G RF technologies' lower latency transmission capabilities. 5G will, for instance, make robotic surgery possible by enabling high-definition video streaming and real-time control in the healthcare industry. On the other hand, maintenance and rescue operations will be significantly safer when using remotely controlled robots in risky industrial settings or catastrophe areas!

3.7.2   Smart Cities

Smart cities provide a larger, more connected cyber-physical environment to live in by integrating IoT into city infrastructure like traffic lights, street lighting, and trash management. As a result, data-driven decisions can be used to optimize operations and cut expenses associated with, for instance, the environmental effects of idling traffic. The interconnection of smart cities will be more common than ever with 5G as a standard for wireless connectivity across mobile phones, PCs, and IoT devices!

3.7.3   Agriculture

Precision farming, which involves precise monitoring and control of a wide range of environmental factors, such as temperature, light levels, air composition, and water usage, is made possible by the use of IoT in agriculture. Asset monitoring and smart irrigation are two further agricultural applications that might be used to manage vast tracts of agriculture from a single central location. Precision farming is now achievable on a far bigger scale because 5G's high speeds make it possible to monitor the health of individual animals in real time, for example.

3.8   RF PRODUCT RECOMMENDATIONS OF IoT

Although the global rollout of 5G infrastructure [9] is still in progress, the next products will let us create robust and efficient IoT solutions that will last for years to come while waiting for general adoption!

3.8.1   Dragino NB-IoT Shield-B5

An Arduino extension board for well-known Arduino boards like the Arduino Uno that supports NB-IoT connectivity is called the Dragino NB-IoT Shield-B5. The Dragino NB-IoT Shield-B5 enables you to create a variety of IoT applications with Arduino, such as smart metering or property management services, security or safety systems, and more, due to its low power consumption and large coverage area.

3.8.2   NB-IoT

A large variety of new IoT devices and services are made possible by the standards-based low power wide area (LPWA) technology known as narrowband-IoT (NB-IoT). "In deep coverage, NB-IoT dramatically increases spectrum efficiency, system capacity, and user device power consumption. A variety of use cases can accommodate a battery life of more than 10 years.

"The challenging requirements of expanded coverage—rural and deep interiors—and ultralow device complexity are met by new physical layer signals and channels. The NB-IoT module's initial cost is anticipated to be similar to that of GSM/GPRS. Although the underlying technology is considerably simpler than GSM/GPRS today, its cost is anticipated to drop significantly as demand rises" [10].

Aimed at students without a background in electronics or programming, Arduino is an open-source electronics platform built on simple-to-use hardware and software. The Arduino board started evolving as soon as it gained a larger audience, diversifying its offering from basic 8-bit boards to items for IoT applications, wearable technology, 3-D printing, and embedded environments. All Arduino boards are fully open-source, enabling users to construct them on their own and eventually customize them to suit their own needs. The program is open-source as well, and users from all over the world are contributing to its growth.

An expansion board for Arduino that adds NB-IoT technology is called the NB-IoT Shield. Users may quickly analyze, evaluate, and execute proof of concept (POC) for NB-IoT solution with NB-IoT Shield and Arduino.

Features include [10]:

- Supports 850-MHz NB-IoT bands;
- Low power consumption;
- Wide area coverage;
- Attention (AT) command to control;
- Autosupport 3.3v or 5v Arduino board;
- Compatible with Arduino Leonardo, Uno, Mega2560, DUE, and more.

Applications include:

- Facility management services, including smart meters for energy, gas, and water;
- Connected personal appliances that measure health indicators;
- Fire and intrusion alarms for residential and commercial sites;
- Smart city infrastructure, such as street lamps or trash cans.

3.8.3 SenseCAP Sensor Hub 4G Data Logger

An industrial-grade, 4G cellular Sensor Hub that is simple to deploy is called the SenseCAP Sensor Hub 4G Data Logger. It comes with 4G for remote connection and communicates with many sensors using the Modbus-RTU RS485 protocol! Solutions for smart cities, agriculture, and environmental monitoring work really well with the SenseCAP Sensor Hub.

The following features are supported:

- Simultaneous collection of multiple environmental data;
- Local data storage;
- Use of standard MODBUS-RTU RS485 sensors;
- Uploading of data to any server (SenseCAP server or user's private server);
- Support for 4G/3G/2G communication;

- Support for the global LTE frequency band;
- Support for remote upgrade and maintenance;
- Built-in GPS location function;
- Two power supply options available: DC only, solar power (to be purchased separately, coming soon);
- Easy to install and deploy, without requirements of engineering background;
- Industry standards, suitable for harsh outdoor environments.

Powerful 4G data logger SenseCAP Sensor Hub can connect to up to 32 RS-485 sensors. It is stable and strong, constructed in accordance with industrial requirements. It is ideal for long-term remote environmental monitoring in outdoor application settings because it is IP66 certified, waterproof, and dustproof.

There are two variations of the Sensor Hub:

Version 1 (DC only): Compatible with solar systems. It should be connected to the entire solar system, not simply a solar panel;

Version 2: A variant with a built-in rechargeable battery that may be utilized with solar power or DC power (will be coming soon).

An easy-to-install 4G cellular station is Sensor Hub Data Logger. It can gather multiple types of sensor data simultaneously and communicates with sensors using the MODBUS-RTU RS485 interface.

There are four RS485 data channels in the Sensor Hub. It can communicate with up to 32 sensors at once via extension hubs or RS485 splitters. Through the use of 4G/3G/2G, the data is gathered and delivered to the cloud (either to the SenseCAP server or the client's private server). The IP66-rated and industry-standard Sensor Hub is appropriate for use in outdoor and challenging conditions and is resistant to ultraviolet (UV) rays, rain, dust, and other elements. A global navigation satellite system (GNSS) is included to track location.

Additionally, Sensor Hub has 10 MB of onboard memory to locally store more than 700,000 measurements in case of a poor connection. The large-capacity lithium battery enables the devices to function for up to two weeks after a power loss or during inclement weather for the version with a built-in rechargeable battery.

The deployment and setup of Sensor Hub just takes a few minutes because of its simple design. It can be mounted on a wall or a pole.

Any sensors that support the MODBUS-RTU RS485 protocols can be used with the SenseCAP Sensor Hub 4G Data Logger. Additionally, a large assortment of industrial-grade RS485 sensors is offered for your consideration. All of these sensors have watertight aircraft connectors, so all you have to do to get them to function is plug them into the Sensor Hub.

Applications include:

- Smart agriculture;
- Smart city;
- Smart buildings;
- Smart industry;
- Environmental monitoring;
- Other wireless sensing applications.

3.8.4   Wio-E5 Development Kit

In the future IoT era, LR will be one of the vital technologies that accompany 5G as it develops [11]. The Wio-E5 Development Board, an antenna, a Universal Serial Bus (USB) type C cable, and a 2*AA 3V battery holder make up the Wio-E5 Dev Kit. The Wio-E5 STM32WLE-5JC module, compatible with LoRaWAN on the global frequency band, is built into the Wio-E5 Dev Board. Additionally, it supports a variety of data interfaces and protocols, including full general-purpose input/output (GPIOs) and RS-485.

Features of full GPIOs led to a variety of rich interfaces, including RS-485, Grove, and others. Other features include support for the global long-range frequency plan, long-distance transmission range to 10 km, ultralow power consumption, and good performance (ideal value in open area).

The Wio-E5 Development Kit is a convenient, small development toolkit that enables you to access the Wio-E5 STM32WLE5JC's potent capabilities. It includes a Wio-E5 Dev Board, an antenna (EU868/US915), a USB type C connector, and a battery holder for two AA batteries at three volts.

The Wio-E5 Dev Board includes the Wio-E5 STM32WLE5JC module, a long-range RF, and microcontroller unit (MCU) chip combination that is a global first, and is (FCC) and CE certified. It is driven by an ARM Cortex-M4 core and a Semtech SX126X long-range chip, and it supports (G)FSK, (G)BPSK, (G)MSK, and long-range modulations in addition to the long-range protocol on the global frequency.

An ultralong transmitting range, incredibly low chip power consumption, and user-friendly interfaces are all aspects of the Wio-E5 Dev Board.

In an open region, the Wio-E5 Dev Board's long-distance transmission range is up to 10 km. The Wio-E5 modules on board have a sleep current of as little as 2.1 uA. (WOR mode). It has a wide operating temperature range of −40°C to 85°C, a high sensitivity range of −116.5 dBm to −136 dBm, and a power output of up to +20.8 dBm at 3.3V.

The Wio-E5 Dev Board features rich interfaces as well. The Wio-E5 Dev Board was created to enable the Wio-E5 module's full functionality. It has led out all 28 of its pins and offers a variety of rich interfaces, such as Grove connectors, RS-485 terminals, and male/female pin headers, allowing you to connect sensors and modules with various connectors and data protocols while saving time on wire soldering. When an external power supply is unavailable, you can still easily power the board by connecting the battery holder with 2 AA batteries. It is an easy-to-use board for quick prototyping and testing.

There are three basic methods to use the Wio-E5 Dev Board because it is a long-range chip with an MCU:

1. Use AT instructions to control the Wio-E5 Dev Board when connected via USB to a PC.

   You could simply connect the Wio-E5 Dev Board to your PC with a USB type C cable and use serial communication software to transmit AT commands and retrieve data from the board because the board has an integrated USB to Universal Asynchronous Receiver/Transmitter (UART) feature.

2. Use UART to connect the Wio-E5 Dev Board to a different mainboard and control it with AT commands.

For instance, attach the Wio-E5 Dev Board to the Seeeduino Xtremely Intelligent Arduino-compatible Object (XIAO) and the Expansion Board using UART, and use the Arduino integrated development environment (IDE) serial monitor to transmit AT commands and read data from the Seeeduino XIAO.

3. Create user applications via the software development kit (SDK).

   Utilizing the STM32Cube Programmer, an official STMicroelectronics SDK, you may create your own long-range development board with MCU functionality. Please check the materials in learning and document to obtain this SDK resource.

The Wio-E5 Dev Board will be a better option for IoT device development, testing, prototyping, and applications in long-distance, ultralow power consumption IoT scenarios including smart agriculture, smart offices, and smart industry thanks to all the amazing capabilities mentioned in Section 3.8.

Applications include:

• Wio-E5 module easy testing;
• Rapid prototyping of long-range devices using Wio-E5;
• Development of any long-distance wireless communication applications;
• Learning and researching long-distance applications.

3.9 CONCLUSION

The IoT is rapidly transforming the world we live in. By enabling devices to connect and communicate with each other, it has the potential to revolutionize industries ranging from healthcare to agriculture. One critical aspect of IoT is RF and microwave technologies. These technologies are essential for enabling communication between IoT devices and ensuring that they function correctly. This chapter explored various aspects of RF and microwave technologies in IoT. It provided an introduction to the topic and discussed the importance of RF and microwave technologies in IoT. It also explored the scope of RF technology in IoT and looks at its applications in various industries, such as wearable technology, smart cities, industrial and automobile, supply chain and retail management, healthcare, energy management, and smart farming and agriculture.

Furthermore, the chapter delved into the fundamentals of RF for IoT, such as sensitivity, transmitter output power, transmitter loss, and antenna gain. It also provided a hypothetical example of RF technology in IoT to illustrate its importance in real-life scenarios. Additionally, it discussed 5G RF technology and how it can transform IoT, as well as its use cases in remote robotics, smart cities, and agriculture. Finally, the chapter provided RF product recommendations for IoT, such as Dragino NB-IoT Shield-B5 and NB-IoT, and looked at their applications in IoT.

Overall, RF and microwave technologies are critical for enabling IoT devices to communicate and function properly. They play a vital role in various industries, such as healthcare, agriculture, and supply chain management. As IoT continues to evolve, the importance of RF and microwave technologies will only increase, and we can expect to see continued innovation and development in this field.

In conclusion, the scope of RF technology in IoT is vast and varied. From wearable technology to smart cities to agriculture, RF and microwave technologies are transforming how we live and work. As the world becomes increasingly connected, the importance of RF and microwave technologies in IoT cannot be overstated. Therefore, it is essential to continue researching and developing these technologies to ensure that IoT continues to evolve and revolutionize the way we live and work.

References

[1] Andreasson, K., and R. Wallace, *Introduction to RF and Microwave Passive Components*, Norwood, MA: Artech House, 2015.

[2] Larson, L., "RF and Microwave Technology Challenges for Internet-of-Things Applications," in *2015 IEEE 15th Topical Meeting on Silicon Monolithic Integrated Circuits in RF Systems*, San Diego, CA, 2015, pp. 61–62, doi: 10.1109/SIRF.2015.7119875.

[3] Wang, X., X. Wang, and S. Mao, "RF Sensing in the Internet of Things: A General Deep Learning Framework," in *IEEE Communications Magazine*, Vol. 56, No. 9, Sept. 2018, pp. 62–67, doi: 10.1109/MCOM.2018.1701277.

[4] Farooq, M. O., and T. Kunz, "IoT-RF: A Routing Framework for the Internet of Things," in *2017 IEEE 28th Annual International Symposium on Per-

sonal, Indoor, and Mobile Radio Communications (PIMRC), Montreal, QC, Canada, 2017, pp. 1–7, doi: 10.1109/PIMRC.2017.8292730.

[5] Jia, X., Q. Feng, T. Fan, and Q. Lei, "RFID Technology and its Applications in Internet of Things (IoT)," in *2012 2nd International Conference on Consumer Electronics, Communications and Networks (CECNet)*, Yichang, China, 2012, pp. 1282–1285, doi: 10.1109/CECNet.2012.6201508.

[6] Mezzanotte, P., V. Palazzi, F. Alimenti, and L. Roselli, "Innovative RFID Sensors for Internet of Things Applications," *IEEE Journal of Microwaves*, Jan. 2021, pp. 55–65, doi: 10.1109/JMW.2020.3035020.

[7] Varsier, N., L.-A. Dufrène, M. Dumay, Q. Lampin, and J. Schwoerer, "A 5G New Radio for Balanced and Mixed IoT Use Cases: Challenges and Key Enablers in FR1 Band," *IEEE Communications Magazine*, Vol. 59, No. 4, April 2021, pp. 82–87, doi: 10.1109/MCOM.001.2000660.

[8] Ejaz W., et al., "Internet of Things (IoT) in 5G Wireless Communications," *IEEE Access*, Vol. 4, 2016, pp. 10310–10314, doi: 10.1109/ACCESS.2016.2646120.

[9] Xu, W., et al., "Opportunities, Challenges and Feasibilities of Zero-Power IoT in 5G Advanced," in *2021 IEEE/CIC International Conference on Communications in China (ICCC Workshops)*, Xiamen, China, 2021, pp. 374–378, doi: 10.1109/ICCCWorkshops52231.2021.9538925.

[10] Nawara, D., and R. Kashef, "IoT-based Recommendation Systems—An Overview," in *2020 IEEE International IoT, Electronics and Mechatronics Conference (IEMTRONICS)*, Vancouver, BC, Canada, 2020, pp. 1–7, doi: 10.1109/IEMTRONICS51293.2020.9216391.

[11] "IEEE Guide for EMF Exposure Assessment of Internet of Things (IoT) Technologies and Devices," *in IEEE Std 1528.7-2020*, Jan. 11, 2021, pp. 1–90, doi: 10.1109/IEEESTD.2021.9319817.

4

BASIC ANTENNAS IN IoT

4.1   INTRODUCTION

IoT applications, particularly those where wired access is virtually impossible, heavily rely on wireless connectivity to communicate with IoT gateways and other devices in the ecosystem. A variety of antennas that support different sorts of networks enable this communication. IoT systems have drastically changed during the last 10 years, becoming smaller and incorporating cutting-edge wireless technologies. These innovations have had a significant impact on the advancement of antenna technology and IoT antenna designs [1], leading to the creation of ultracompact antennas with excellent performance and efficiency. It has become a usual requirement for high-performance, compact form–factor IoT designs to incorporate many antennas, which presents considerable difficulties for IoT product developers. Antenna system design is given even more importance as IoT ecosystems attempt to support high-density, low-latency networks and include new functionalities into radios and overall system topologies. Because of this, engineers see antennas, whether external or integrated, as essential components of IoT applications and the development of smart surroundings rather than viewing them as passive goods.

4.2 DESIGN OF ANTENNAS

When a coordinate system is centered on an antenna, spherical waves it emits radiate and move in the radial direction. Spherical waves can be roughly modeled by plane waves at great distances. Plane waves are advantageous since they make the issue simpler. However, because they demand infinite power, they are not corporeal. All of these are well mathematically explained in [2]. The following equations throughout this chapter detail the same.

The electromagnetic wave power density and propagation direction are both indicated by the Poynting vector. It is identified by the electric (E) and magnetic field (H) vector cross product, and it is designated S as in:

$$S = E \times H \quad W/m^2 \tag{4.1}$$

The vector represents the rate at which energy is flowing through a unit area perpendicular to the direction of wave propagation. The direction of the Poynting vector is perpendicular to both the electric field and magnetic field vectors, and is determined by the right-hand rule.

The magnitude of the Poynting vector at any point in space gives the power per unit area being carried by the electromagnetic wave at that point. For a plane wave, the magnitude of the Poynting vector is given by:

$$|S| = (1/2) * \varepsilon* E^2 \tag{4.2}$$

Where ε is the permittivity of the medium through which the wave is propagating, and E is the magnitude of the electric field.

The Poynting vector is a fundamental concept in electromagnetism, and is used in a wide range of applications, including antenna theory, optics, and particle physics. In antenna theory, the Poynting vector is used to analyze the power transfer from the antenna to the surrounding environment [3], and to calculate the radiation pattern and efficiency of the antenna. In optics, the Poynting vector is used to describe the flow of light energy in optical systems, and to calculate the intensity and direction of light propagation. In particle physics, the Poynting vector is used to describe the flow of energy in particle beams and in interactions between particles and electromagnetic fields.

4.2.1 Need for Matching

A resonant device is one that produces more power within a small range of frequencies. Such resonant devices include antennas, whose outputs are improved if the impedance is matched. If the antenna impedance matches [4] the free space impedance, the power radiated by the antenna will be efficiently radiated. For a receiver antenna, the output impedance of the antenna must match the input impedance of the receiver amplifier circuit; for a transmitter antenna, the input impedance of the antenna must match the output impedance of the transmitter amplifier as well as the transmission line impedance.

Matching [5] involves adjusting the impedance of the antenna to match the impedance of the transmission line to which it is connected. The goal of matching is to minimize the amount of power that is reflected back toward the source and maximize the amount of power that is transferred to the antenna.

Mathematically, the impedance of an antenna is represented by a complex number that has both a magnitude (or resistance) and a phase angle. The impedance of the transmission line is also represented by a complex number that has both a magnitude and a phase angle. In order to match the impedance of the antenna to the transmission line, the magnitude and phase angle of the two complex numbers must be equal.

If the impedance of the antenna is not properly matched to the transmission line, some of the energy that is sent from the source to the antenna will be reflected back toward the source, rather than being transferred to the antenna. This can cause a reduction in the amount of power that is available to the antenna, and can result in poor performance or damage to the source or transmission line.

Mathematically, the amount of power that is reflected back toward the source due to a mismatched impedance can be calculated using the reflection coefficient (Γ). The reflection coefficient is given by:

$$\Gamma = (ZL - Z0) \, / \, (ZL + Z0) \qquad (4.3)$$

where ZL is the impedance of the load (i.e., the antenna), $Z0$ is characteristic impedance of the transmission line, and Γ is a complex number that represents the amount of power that is reflected back toward
the source.

The magnitude of the reflection coefficient can be used to determine the amount of power that is reflected back toward the source. If
the magnitude of the reflection coefficient is high, then a significant
amount of power is being reflected back toward the source, indicating
a poor impedance match between the antenna and transmission line.

Matching the impedance of the antenna to the transmission line
is therefore necessary to ensure maximum power transfer from the
source to the antenna, and to prevent damage to the source or transmission line. This is achieved by adjusting the resistance and reactance
components of the antenna and transmission line, typically using a
matching network or impedance transformer. By properly matching
the impedance of the antenna to the transmission line, the amount of
power that is reflected back toward the source can be minimized, and
the efficiency and performance of the antenna can be maximized.

4.2.2 Matching Impedance

The usual definition [5] states that impedance matching occurs when
the approximate values of the impedances of a transmitter and receiver are equal or when they are opposite. Between the electronics and
the antenna, impedance matching is required. In order to maximize
power transfer between the antenna and the receiver or transmitter,
the electronics, transmission line, and antenna impedances should
all be compatible.

The mathematical expression for the impedance of an antenna is
given as follows:

$$Za = Ra + jXa \tag{4.4}$$

where Za is the impedance of the antenna, Ra is the resistance component of the impedance, and Xa is the reactance component of the
impedance.

Similarly, the mathematical expression for the impedance of a
transmission line is seen:

$$Zt = Rt + jXt \tag{4.5}$$

where Zt is the impedance of the transmission line, Rt is the resistance component of the impedance, and Xt is the reactance component of the impedance.

To match the impedance of the antenna to the transmission line, the complex impedance of the antenna must be equal to the complex impedance of the transmission line. This can be expressed mathematically as:

$$Za = Zt \tag{4.6}$$

or

$$Ra + jXa = Rt + jXt \tag{4.7}$$

To solve this equation, the resistance and reactance components of the antenna and transmission line must be adjusted. One common method of matching impedance is to use a matching network, which is a network of components that is placed between the antenna and the transmission line to adjust the impedance of the system.

The matching network can be designed using various techniques, such as the Smith chart, which is a graphical tool for solving transmission line problems. The Smith chart allows the designer to visualize the complex impedance of the antenna and transmission line and to select the appropriate matching network components to achieve a match.

4.2.3 Reflected Power and VSWR

The VSWR, as stated in the standard definition [2], is the ratio of the greatest voltage to the minimum voltage in a standing wave. The power cannot be properly disseminated if the circuitry, transmission line, and antenna impedances are not compatible. The power is partially reflected back instead. The phrase VSWR denotes the impedance mismatch. Voltage standing wave ratio is abbreviated as VSWR [6]. SWR is another name for it. The value of VSWR will increase with increasing impedance mismatch. For effective radiation, the VSWR should ideally be 1:1. Wasted forward power is referred to as reflected power. Reflected power and VSWR both point to the same conclusion.

VSWR is a measure of the quality of impedance matching between an antenna and its transmission line. It is a dimensionless quantity that describes the ratio of the maximum voltage to the minimum voltage on a transmission line.

Physically, VSWR is caused by the reflection of electromagnetic waves at the point where the transmission line meets the antenna. When there is a mismatch between the impedance of the antenna and the transmission line, a portion of the signal is reflected back toward the source. This reflected signal can interfere with the transmitted signal and reduce the efficiency of the antenna system.

Mathematically, VSWR is defined as the ratio of the maximum voltage amplitude to the minimum voltage amplitude along the transmission line. It can be calculated using the following formula:

$$VSWR = (Vmax\ /\ Vmin) \tag{4.8}$$

where $Vmax$ is the maximum voltage amplitude on the transmission line, and $Vmin$ is the minimum voltage amplitude on the transmission line. In practice, VSWR is often measured using specialized test equipment such as a network analyzer. The measurement is typically represented as a graph of VSWR versus frequency. The graph shows how VSWR varies with frequency and provides a visual representation of the quality of impedance matching between the antenna and transmission line. A low VSWR (typically less than 2:1) indicates that the impedance of the antenna is closely matched to the impedance of the transmission line, while a high VSWR (typically greater than 3:1) indicates that there is a significant mismatch between the two impedances.

The VSWR of an antenna is an important parameter in antenna design, as it affects the performance of the antenna system. A high VSWR can lead to reduced efficiency, increased signal loss, and increased interference. Therefore, antenna designers must carefully match the impedance of the antenna to the transmission line to ensure that the VSWR is within acceptable limits.

4.2.4   Bandwidth

A band of frequencies in a wavelength, designated for the particular transmission, is known as bandwidth, according to the standard definition [2].The signal is spread across a number of frequencies when it is sent or received. This specific frequency range is reserved for a certain signal so that other signals would not interfere with it while it is being transmitted. The range of frequencies between higher and lower frequencies over which a signal is carried is known as the bandwidth [7]. Once allocated, the bandwidth is unavailable to other users. To allocate different transmitters, the entire spectrum is divided into bandwidths. This bandwidth is also known as absolute bandwidth.

Bandwidth is a measure of the range of frequencies over which an antenna can effectively transmit or receive electromagnetic waves. The bandwidth of an antenna is defined as the frequency range over which the antenna's input impedance remains within a specified range, typically within 90% of its nominal value.

The bandwidth of an antenna can be calculated using the following:

$$BW = f2 - f1 \qquad (4.9)$$

where BW is the bandwidth of the antenna, $f1$ is the lower cutoff frequency of the antenna, and $f2$ is the upper cutoff frequency of the antenna.

The lower cutoff frequency of the antenna is the frequency at which the input impedance of the antenna drops below the specified range, and the upper cutoff frequency is the frequency at which the input impedance of the antenna exceeds the specified range.

The input impedance of an antenna is the impedance seen by the transmitter or receiver when it is connected to the antenna. It is a function of the frequency of the electromagnetic wave and is influenced by the geometry and materials of the antenna.

The bandwidth of an antenna is an important parameter in antenna design, as it determines the range of frequencies over which the antenna can operate effectively. A wider bandwidth indicates that the antenna can operate over a broader range of frequencies, while a narrower bandwidth limits the frequency range of operation.

In practical antenna design, the bandwidth is often a compromise between several competing factors, including antenna size, radiation pattern, and efficiency. Designers may use techniques such as adding parasitic elements, adjusting the length or spacing of elements, or changing the shape of the antenna to achieve the desired bandwidth while maintaining other design criteria.

It should be noted that the bandwidth of an antenna is a key factor in determining its performance in real-world applications. Antennas with narrow bandwidths may be less suitable for applications such as communication systems, where the signal frequencies may vary widely, while antennas with wider bandwidths may be better suited to applications such as radio astronomy, where a broad frequency range is required to observe different celestial objects.

4.2.5 Percentage Bandwidth

The standard definition [2] states that percentage bandwidth can be defined as the ratio of absolute bandwidth to the center frequency of that bandwidth.

Resonant frequency is the specific frequency within a frequency band where the signal is strongest. It is also known as the band's central frequency (f_c).

- The letters f_h and f_l, respectively, stand for the higher and lower frequencies.
- The formula $f_h - fl$ yields absolute bandwidth.

Either fractional bandwidth [8] or percentage bandwidth calculations must be made to determine how broad the bandwidth is.

4.2.6 Radiation Levels

Radiation emitted from an antenna that is more intense in a specific direction shows the maximum intensity of that antenna. Radiation intensity is defined as the power per unit solid angle [2]. Radiation intensity is nothing more than the emission of radiation to the greatest extent achievable. Watts/steradian or watts/radian² is the unit used to measure radiation intensity. The focus of the beam and its effectiveness in that direction have a direct impact on the radiation intensity of an antenna. In simple terms, these levels refer to the amount of electromagnetic energy that an antenna radiates into the surrounding

space. These levels are typically measured in terms of power density, which is the amount of electromagnetic energy per unit area. The
power density can be expressed mathematically as:

$$S = P / \left(4 * \pi * r^2\right)$$ (4.10)

where S is the power density, P is the radiated power of the antenna,
and r is the distance from the antenna.

The radiated power of an antenna is the total power that the antenna radiates in all directions. It can be calculated by integrating
the radiation intensity of the antenna over all directions. The radiation intensity of the antenna is the power per unit area per unit solid
angle, and it can be expressed mathematically as:

$$D(\theta, \phi) = r^2 * S(\theta, \phi) / P$$ (4.11)

where $D(\theta, \phi)$ is the radiation intensity of the antenna in the direction
of angles θ and ϕ, $S(\theta,\phi)$ is the power density of the antenna in the
direction of angles θ and ϕ, and P is the radiated power of the antenna.

The radiation levels of an antenna are important considerations
in many applications, including telecommunications, radio broadcasting, and radar. In order to ensure that radiation levels are within
acceptable limits, it is often necessary to model the radiation pattern
of the antenna and calculate the power density at various distances
from the antenna. This can be done using numerical methods such as
the method of moments (MoM) or the finite element method.

It should be noted that radiation levels can have both positive
and negative effects. On the one hand, they are essential for many
communication and sensing applications. On the other hand, high
levels of electromagnetic radiation can be hazardous to human health
and the environment. Therefore, it is important to carefully consider
the radiation levels associated with any antenna installation, and to
take appropriate measures to minimize any potential risks.

4.2.7 Directivity

The conventional definition [2] states that the directivity is defined as
the ratio of the maximum radiation intensity of the subject antenna to
the maximum radiation intensity of an isotropic or reference antenna,
emitting the same total power. An antenna emits power, but the di-

rection of that power is very important. The antenna being watched is referred to as the subject antenna. While sending or receiving, its radiation strength is concentrated in that particular direction. The antenna is said to have that specific direction of directivity as a result.

The term directivity refers to the ratio of an antenna's radiation intensity in one direction to its radiation intensity averaged across all directions, or, if no specific direction is given, the direction in which maximum intensity is observed. The directivity [8] of a nonisotropic antenna is equal to the ratio of its radiation intensity in one direction to its isotropic source antenna. The directivity of an antenna can be expressed mathematically as:

$$D = 4 * \pi \ / \ Pmax \tag{4.12}$$

where D is the directivity of the antenna, and $Pmax$ is the maximum radiation intensity of the antenna.

The maximum radiation intensity of the antenna is the highest radiation intensity that occurs in any direction in the radiation pattern of the antenna. It can be expressed as:

$$Pmax = max(D(\theta,\phi) \tag{4.13}$$

where $D(\theta, \phi)$ is the radiation intensity of the antenna as a function of the angles θ and ϕ.

The directivity of an antenna is a measure of how well the antenna focuses its radiation on a particular direction. A higher directivity indicates that the antenna is better at concentrating its radiation in a particular direction, while a lower directivity indicates that the antenna radiates more uniformly in all directions.

The directivity of an antenna is an important parameter in antenna design, as it determines the gain of the antenna. The gain of an antenna is the measure of how much the antenna amplifies the input signal. It is related to the directivity by:

$$G = D * Ae \ / \ \lambda^2 \tag{4.14}$$

where G is the gain of the antenna, Ae is the effective aperture of the antenna, and λ is the wavelength of the electromagnetic radiation.

It should be noted that the directivity is an idealized parameter that assumes that the antenna radiates uniformly in all directions except for the direction of maximum radiation. In practice, real antennas may exhibit some deviations from this ideal behavior due to factors such as losses, feedline effects, and mutual coupling between antennas in an array.

4.2.8 Aperture Efficiency

The ratio of the effective radiating area (or effective area) to the physical area of the aperture is what is meant by the standard definition of aperture efficiency of an antenna [2]. The aperture of an antenna is where the power is transmitted. This radiation ought to be efficient with little loss. The physical area of the aperture should also be taken into account because it affects the radiation's efficacy and is physically located on the antenna.

The effective aperture of an antenna is the area of an ideal aperture that would produce the same power as the actual antenna for a given incident power density. It can be expressed mathematically as:

$$Ae = \left(\lambda^2 / 4 * \pi \right) * \iint D(\theta, \phi) * \sin(\theta) \ d\theta \ d\phi \qquad (4.15)$$

where Ae is the effective aperture of the antenna, $D(\theta, \phi)$ is the radiation intensity of the antenna as a function of the angles θ and ϕ, and the integrals are taken over the entire solid angle.

The physical aperture of an antenna is the actual area of the antenna that is available to capture or radiate electromagnetic energy. It can be expressed as:

$$Ap = \pi * \left(D / 2 \right)^2 \qquad (4.16)$$

where Ap is the physical aperture of the antenna, and D is the diameter of the antenna.

The aperture efficiency of an antenna is then given by the ratio of the effective aperture to the physical aperture as:

$$Eff = Ae \ / \ Ap \qquad (4.17)$$

where Eff is the aperture efficiency.

The aperture efficiency of an antenna is an important factor in the overall performance of a communication system. A more efficient antenna will be able to capture or radiate more power over a given area, resulting in a stronger signal and better communication quality. Conversely, a less efficient antenna will be able to capture or radiate less power, resulting in a weaker signal and poorer communication quality.

It should be noted that the aperture efficiency is a function of the radiation pattern of the antenna, which is typically measured in the far field of the antenna. The radiation pattern of the antenna describes how the antenna radiates energy in space and is affected by the size and shape of the antenna, as well as the frequency of operation. Therefore, to accurately determine the aperture efficiency of an antenna, it is important to measure its radiation pattern in the far field.

4.2.9 Antenna Efficiency

The conventional definition [2] states that antenna efficiency is the ratio of the antenna's output power to its input power accepted. Simply put, an antenna's purpose is to radiate power from its input with the fewest possible losses. The effectiveness of an antenna describes how well it can deliver its output with the fewest transmission line losses. This is also known as the antenna's radiation efficiency factor. The efficiency of an antenna can be expressed mathematically as:

$$\eta = Pr \: / \: Pi \qquad (4.18)$$

where η is the antenna efficiency, Pr is the radiated power, and Pi is the input power.

The input power to an antenna can be expressed in terms of the voltage and current at the input terminals of the antenna as:

$$Pi = 0.5 \: {}^* \: Re\{VI\} \qquad (4.19)$$

where $Re\{\}$ denotes the real part of the complex quantity inside the brackets, V is the voltage at the input terminals of the antenna, and I^* is the complex conjugate of the current at the input terminals of the antenna.

The radiated power from an antenna can be calculated using the radiation resistance of the antenna and the current flowing through the antenna as:

$$Pr = 0.5 * Re\{ZrI\} \qquad (4.20)$$

where Zr is the radiation resistance of the antenna, which represents the resistance that would be required in a hypothetical resistor to dissipate the same amount of power as that radiated by the antenna, and I^* is the complex conjugate of the current flowing through the antenna.

The radiation resistance of an antenna can be calculated using the antenna's geometry and the wavelength of the electromagnetic energy it is radiating or receiving. For example, the radiation resistance of a dipole antenna can be expressed as:

$$Zr = \left(2 * \pi * f * L\right)^2 / \left(4 * \pi * c\right) \qquad (4.21)$$

where f is the frequency of operation, L is the length of the dipole, and c is the speed of light in a vacuum.

The efficiency of an antenna is an important factor in the overall performance of a communication system. A more efficient antenna will radiate more power for a given input power, resulting in a stronger signal and better communication quality. Conversely, a less efficient antenna will waste more of the input power as heat and will radiate less power, resulting in a weaker signal and poorer communication quality.

4.2.10 Gain

Gain is defined as the ratio of the radiation intensity in a given direction to the radiation intensity that would be obtained if the power accepted by the antenna were emitted isotropically, in accordance with the standard definition [2]. Simply said, an antenna's gain accounts for both its effective performance and directivity. The radiation intensity received can be used as a reference of whether the power the antenna accepted was radiated isotropically, or in all directions.

Antenna gain refers to the amount of energy transmitted from an isotropic source to the direction of peak radiation. The gain of an antenna can be expressed in decibels as in:

$$G = 10 \; log \left(P\_r \; / \; P \quad i\right) \tag{4.22}$$

where G is the gain of the antenna in dB, P_r is the power radiated or received in a particular direction, and P_i is the power that would be radiated or received by an isotropic radiator radiating or receiving the same total power.

Alternatively, the gain can be expressed as a linear ratio as in:

$$G = P\_r \; / \; P\_i \tag{4.23}$$

where G is the gain of the antenna, P_r is the power radiated or received in a particular direction, and P_i is the power that would be radiated or received by an isotropic radiator radiating or receiving the same total power. In contrast to directivity, antenna gain considers all losses and so focuses on efficiency. The gain of an antenna is often specified in terms of the antenna's maximum gain, which is the maximum gain observed in any direction. The maximum gain is typically achieved when the antenna is pointed directly at the source or destination of the electromagnetic energy. The maximum gain of an antenna can be calculated using the antenna's radiation pattern and the power that would be radiated or received by an isotropic radiator radiating or receiving the same total power. The near-field and far-field areas of the antenna are a crucial topic to take into account after the antenna parameters. When measured closer to the antenna, the radiation intensity is different from when it is measured farther away from the antenna. Despite being far from the antenna, the location is nevertheless thought to be effective because of the high radiation intensity.

4.2.11 Near Field

Near field refers to the field that is closer to the antenna. Despite having certain radiation components, it has an inductive effect, which is why it is also known as an inductive field. In the context of antennas, the near field [9] is the region close to the antenna where the electromagnetic fields are predominantly reactive, and the behavior of the fields cannot be described by the radiating wave. The near field can be further divided into two regions: the reactive near field and the radiating near field.

The reactive near field is the region immediately adjacent to the antenna, where the electromagnetic fields are primarily stored as electric and magnetic energy. The fields in this region decay rapidly with distance from the antenna, and their behavior can be described using the equations of electrostatics and magnetostatics.

The radiating near field is the region beyond the reactive near field but still close to the antenna, where the electromagnetic fields are predominantly radiating but have not yet reached their far-field behavior. In this region, the fields are a combination of radiated and stored energy, and their behavior is described using the equations of electromagnetics.

The distance at which the near field transitions to the far field is called the Fraunhofer distance or the far-field distance. This distance is given by:

$$D = \left(2 * D^2\right) / \lambda \qquad (4.24)$$

where D is the maximum dimension of the antenna aperture, and λ is the wavelength of the electromagnetic radiation.

In the near field, the fields can be described using the vector potential A and the scalar potential ϕ. The electric and magnetic fields can be expressed as:

$$E = -\nabla\varphi - \partial A / \partial t \qquad (4.25)$$

$$B = \nabla \times A \qquad (4.26)$$

where ∇ is the gradient operator, t is time, and $\times$ represents the cross product.

In summary, the near field of an antenna is the region close to the antenna where the electromagnetic fields are predominantly reactive and cannot be described by the radiating wave. The behavior of the fields in this region can be described using the equations of electrostatics, magnetostatics, and electromagnetics. The transition from the near field to the far field occurs at a distance given by the Fraunhofer distance formula given as:

$$D = \left(2a^2\right) / \lambda \qquad (4.27)$$

where D is the distance from the aperture where the Fraunhofer (far-field) approximation becomes valid, a is the size of the aperture, and λ is the wavelength of the radiation.

4.2.12   Far Field

Far field refers to a field that is away from the antenna. As the radiation effect is strong here, it is also known as a radiation field. Numerous antenna factors, including antenna directivity and radiation pattern, are only taken into account in this area. In the context of antennas, the far field is the region at a distance from the antenna where the electromagnetic fields can be accurately described as a propagating wave. The far field can be further divided into two regions: the Fresnel region and the Fraunhofer region.

The Fresnel region is the region immediately beyond the near field where the wavefronts are still curved, and the fields are a mixture of both reactive and radiative components. In this region, the fields can be described using the Fresnel integrals.

The Fraunhofer region is the region beyond the Fresnel region where the wavefronts are approximately planar, and the fields are purely radiative. In this region, the fields can be described using the far-field approximation, which assumes that the fields are a propagating wave and that the electric and magnetic fields are orthogonal to the direction of propagation. The far-field distance can be approximated by the Fraunhofer distance formula as mentioned in Section 4.2.11.

4.2.13   Field Pattern

Field pattern [10] refers to the field distribution that can be quantified in terms of field intensity. In other words, the plotted radiated power from the antenna is expressed as an electric field, E (v/m). The field pattern of an antenna is a graphical representation of how the antenna radiates energy in space. The field pattern shows the relative strength and direction of the electromagnetic fields at different points in space, and it is a useful tool for understanding the behavior and performance of antennas. There are two types of field patterns: the radiation pattern and the impedance pattern. The radiation pattern describes how the antenna radiates energy in space, while the impedance pattern describes how the antenna impedance varies with position.

4.2.14 Radiation Pattern

The radiation pattern of an antenna is a graphical representation of the relative strength and direction of the electromagnetic fields radiated by the antenna [11]. The radiation pattern is typically presented in a polar or rectangular coordinate system, with the antenna located at the origin.

The radiation pattern [12] is a function of the direction of observation, and it is typically measured in the far field of the antenna. The far-field region is characterized by a nearly uniform wavefront, and the electromagnetic fields can be approximated as a propagating plane wave.

In many cases, the radiation pattern is presented in terms of the directivity of the antenna, which is defined as the ratio of the maximum radiation intensity in a given direction to the average radiation intensity over all directions.

4.2.15 Impedance Pattern

The impedance pattern of an antenna is a graphical representation of how the antenna impedance varies with position. The antenna impedance is a complex quantity that describes the relationship between the voltage and current at the terminals of the antenna.

In general, the antenna impedance varies with position because of the effects of ground planes, nearby structures, and other objects in the environment. The impedance pattern is typically measured in the near field of the antenna, where the electromagnetic fields are predominantly reactive.

4.3 TYPES OF ANTENNAS

4.3.1 Antenna Arrays

An antenna may emit a certain amount of energy in one direction when used alone, which improves transmission. If a few additional elements are added to the antenna, the output will be more efficient. Exactly this concept inspired the development of antenna arrays. A radiating system made up of distinct radiators and elements is an antenna array, as seen in Figure 4.1.

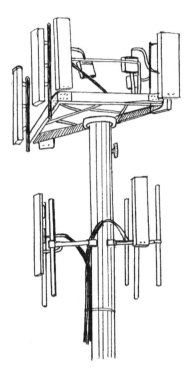

Figure 4.1 Antenna array.

While operating, each of these radiators has its own induction field. The elements are arranged so closely together that their induction fields overlap. Thus, the radiation patterns they would emit would equal the vector sum of each one. When designing these antennas, it is also important to consider the element spacing and element length in relation to wavelength.

Each antenna radiates on its own, but when arranged in an array, all of the elements combine their radiation to create a single, high-gain, high-directivity beam that performs better while generating the fewest losses.

Advantages include:

- Higher signal-to-noise ratio;
- Higher gain;
- Lower power consumption;
- Decreased minor lobes;

- Increased signal intensity;
- High directivity;
- Reduced minor lobes;
- Improved performance.

Disadvantages include:

- Increased resistive losses;
- Challenging mounting and maintenance;
- Huge exterior space requirement.

Applications include:

- Astronomical research;
- Military radar communications;
- Wireless communications;
- Satellite communications;
- Astronomy.

4.3.2 Bow-Tie Antennas

Instead of using straight rods as the antenna elements, a bow-tie antenna [13] makes use of triangular elements, as seen in Figure 4.2.

The name of the antenna comes from the triangular protrusions on either side that resemble a bow-tie. On either side of the guiding beam, the antenna's "wings" flare out symmetrically. At their center,

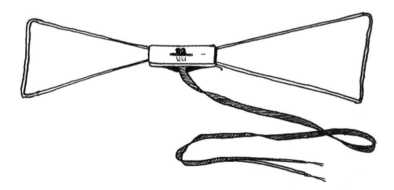

Figure 4.2 Bow-tie antenna.

the two antennas are almost touching. Because of how it resembles a butterfly with its wings held open, a bow-tie antenna is occasionally referred to as a butterfly antenna. A cat's whisker antenna is one in which the bow-tie components contain a metal bar that closes the antenna.

An example of a UHF fan dipole antenna is the bow-tie antenna. Even though bow-tie antennas resemble log periodic ones, they are not regarded as linearly polarized (LP) antennas. Some individuals do view the bow-tie antenna as a more straightforward variant of the log periodic tooth antenna. An example of a biconical antenna is a bow-tie antenna. The genuine biconical antenna is thought of as a 2-D counterpart of the bow-tie antenna. Several pieces protrude from a real biconical antenna in a 360-degree pattern in both directions. The antenna's bandwidth is improved by utilizing triangular elements rather than straight ones, and it frequently receives signals at an angle of 60 degrees, which is ideal for picking up signals from multiple sources. Bow-tie antennas were frequently used with televisions receiving over-the-air broadcasts because of this. They may pick up a variety of UHF frequencies, and will perform significantly better in this regard than a thin wire dipole antenna. The structure of this antenna makes it more wind-resistant and lighter than a fan dipole antenna that joins the horizontal elements. The metal bar that seals the bow-tie makes the design stronger; however, it does add to the weight of the antenna, which is a bonus when you don't want to have to climb on the roof to fix the antenna so you can watch the news. This antenna design is inexpensive and simple to build. Because of this, rabbit ear antennas were frequently found on vintage televisions. Compared to other antenna types, such as Yagi antennas, the mesh reflector of the bow-tie antenna is more effective. In the low end of their frequency spectrum, biconical antennas—including bow-tie antennas—have poor transmitting efficiency. In this case, a log periodic antenna is preferable.

4.3.3 Helical Antennas

Helical antennas [14] are the most basic antennas that are commonly employed in ultrahigh frequencies, hence this antenna operates in both the VHF and UHF bands. These antennas are made of helix-shaped conducting wires as seen in Figure 4.3.

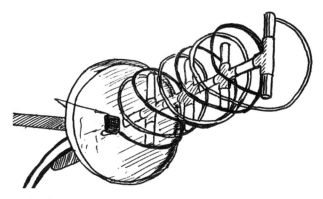

Figure 4.3 Helical antenna.

A few distinctive features of this antenna include its wide bandwidth, high gain, and circular polarization. The following are the key characteristics of a helical antenna:

- This straightforward antenna is employed in circular polarization.
- It is used in a variety of frequencies, including VHF and UHF. It is typically utilized in axial mode.
- In this mode, where efficiency and beamwidth are low, it is not selected.
- Its design is quite straightforward and has high directivity. It functions as a wideband antenna in axial mode. Linear horizontal polarization happens when the axial ratio is zero. The occurrence of linear vertical polarization depends on the axial ratio.
- If the axial ratio is one, then circular polarization occurs.

Advantages are:

- Design is simple;
- Directivity is high;
- Wide bandwidth;
- Circular polarization can be obtained;
- Can be used at VHF and HF bands;
- Robust construction;

• When it uses a circularly polarized pattern then it is acceptable through both vertical and horizontal polarized antenna types.

Disadvantages of a helical antenna include the following:

• Its size is larger, so it occupies more space.
• The efficiency mainly depends on the number of turns so, because of the number of turns, the efficiency will be decreased.
• High cost.

Applications of helical antennas include the following:

• These antennas are applicable in satellite and space probe communications because of their circular polarization of the transmitted electromagnetic waves and maximum directivity;
• A single or array of helical antennas are used for transmitting and receiving VHF signals;
• Used for satellites at Earth stations;
• Used for telemetry links through ballistic missiles;
• Communication can be established between the moon and the Earth;
• Helical antennas are used in many satellites like data relay and weather;
• This antenna is used for transmitting and receiving VHF waves, especially for ionospheric propagation;
• It is used for different communications like radio astronomy, space telemetry, satellite, and space.

PCB Antennas[1]

Compact low power devices have always been preferred, especially in automotive and smart wearables due to the consumer demand for high-tech functionalities in one package. Alongside network compatibility of PCB antennas, their high strength makes them robustly connective and durable, thus embedded in the majority of consumer electronic devices.

---

1. *From:* [15].

PCB antennas, as seen in Figure 4.4, operating at high frequency enable devices to communicate over long distances by transducing the electric signals into electromagnetic waves.

The main advantage of PCB antennas is that they reduce the physical footprint and the maintenance cost of the device. The size of an antenna should be compact and small, and in order to increase the efficiency, several microstrip patches are attached together to acquire the desired gain from the small size. The size of the patches depends directly on the wavelength of the operating frequency.

Antennas are very sensitive to their surroundings; thus, when an antenna is embedded into a PCB, the design and the layout should be considered as per the requirements as this may have a huge impact on the performance of the wireless device. Even minute details like material, layer count, or layer thickness can have an effect on the antenna's performance. When it comes to designing a PCB antenna, there are various steps to be taken into account. Some of them are as follows.

1. Positioning the Antenna

Antennas have different modes of operations and depending on the radiation level of an antenna, there are certain positions where they must be placed. For example, along the short side of the PCB, the long side of the PCB, or in the corner of the PCB. Ideally, the corner of the PCB is one of the optimal places to position an antenna. This is because the corner position of a PCB allows the antenna to have clearance in up to five spatial directions and the feed to the antenna lies in the sixth direction. There are various antenna designs that are best suited for different positions, hence PCB designers can select the antenna according to their application and layout.

Figure 4.4 PCB antenna.

2. Keep-Out Area

Designers must strictly ensure that components must not be placed in the near field directly around the antenna, as this may result in signal interference, which will affect the performance of the circuit. Also, it should be made sure that the area around the antenna must be spaced from metallic objects including mounting screws. The antenna radiates against a ground plane and the ground plane is associated with the frequency at which the antenna operates. Thus, allowing the proper size and spacing for the antenna's ground plane is a must.

3. Ground Planes

The size of the ground plane on a PCB is an important factor to consider, as any wires used for communication to various devices and batteries that provide potential to the device may alter if not designed correctly. Designers have to make sure that the ground planes are properly sized as it ensures cables and batteries connected to the device have less effect on the antenna. There are certain PCB antennas that are ground-plane dependent, which means that the PCB itself becomes the ground section of the antenna to operate in order to balance the antenna currents and lower layers of the PCB, which may affect the antenna's performance. In such cases, designers must make sure that no battery is placed near the antenna.

4. Proximity to Other PCB Components

During designing, it is crucial to keep the embedded antenna at a distance well away from other circuit components which may interfere with the radiation characteristics of the antenna.

4.3.4 Planar Antennas

Low to medium gain characterizes most planar antennas [16]. In radar systems, a reflecting screen may be present behind the active plane of a planar antenna. A monopole antenna grounded at one end and parallel to the ground plane makes up an inverted F antenna. A planar inverted F antenna, often known as a PIF antenna or PIFA, is seen in Figure 4.5.

As the internal antenna in cell phones, this kind of antenna is frequently utilized. This antenna's multiband variant can be applied to wireless communication gear, satellite navigation systems, and automobile radios.

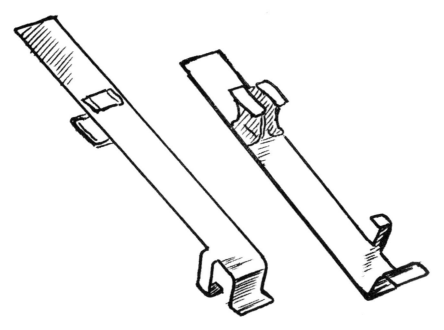

Figure 4.5 PIFA.

Some advantages are:

- Planar antennas are an affordable ultrawideband antenna, particularly when they are manufactured in large quantities using PCB technology.

- They are perfect for wireless applications due to their potential compact size.

- The aperture of planar arrays is big. The phase of each constituent is changed to regulate the direction of the beam.

- Planar antennas are discreet. Since planar antennas on a thin, flexible substrate have no impact on the aerodynamics of the vehicle, they have been employed on aircraft for many years.

- They are strong when mounted on a solid surface. For instance, heat sinks could be incorporated into the planar form.

- Using PCB technology, integrating a planar antenna with additional electronic parts at the same time is neither difficult nor expensive. An integrated planar antenna may be fed directly or indirectly by the transmission lines.

- Depending on the design, they can accommodate both linear and circular polarization.

Disadvantages are:

- The bandwidth of many planar antennas is restricted. When it comes to microstrip antennas, this is very dangerous. They are not very effective at radiating.
- The gain is low. By loading notches and inserting a shorting pin into the radiating patch, this can be improved. The gain of the planar antenna is also increased by the gaps between the patches. Due to this, quad patch antennas with four somewhat smaller-sized patches rather than an antenna with a single, extremely big patch are more common.

Applications include:

- In wireless applications, planar antennas are frequently employed, particularly when the wireless devices may operate on a variety of frequencies. A multiband Wi-Fi network is the standard illustration of this.
- In software defined radio, ultrawideband planar antennas are frequently employed. Spectrum analyzers, cell phone testers, and signal sources can all be used with wideband planar antennas. Radar systems frequently employ planar arrays.

4.3.5 Polymer Antennas

The use of flexible electronics has grown more important for applications that require flexible [17] displays as well as biological applications that have intricate curvilinear geometries. Researchers have conducted experiments using a wide variety of materials, such as polymers, plastic, paper, textiles, and fabrics, in order to increase the adaptability of electrical systems. These materials serve as the substrate for these systems. Such a polymer antenna is seen in Figure 4.6.

Each of these materials possesses its own distinct qualities that determine the degree to which they can be bent, twisted, or crumpled with relative ease. Because of their adaptability and malleability,

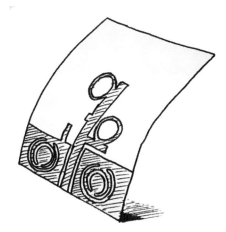

Figure 4.6 Polymer antenna.

these materials are ideal for use in the development of designs for forthcoming intelligent electronic systems, which may also find application in the IoT.

There are a variety of electrical and communication applications that require the flexibility of a variety of materials, including applications as diverse as flexible screens, wearable technology, smart tags, and flexible antenna improvements. In fact, in today's world, flexible screens and antenna systems are indispensable to healthcare, business, the military, and even personal connection. These versatile devices can be utilized in a broad variety of contexts, including RFID tagging, aviation, and health monitoring systems. A lightweight and conformal design option is provided by flexible circuits, which are similar to thin sheets of carbon nanotube on plastic substrates. These flexible integrated circuits offer a wide range of possible applications in embedded systems and other disciplines of electronics that are already familiar with the use of flexible RFID tags of a variety of forms. Recent examples of flexible electronics include stretchable organic solar cells that can be used as biological sensors, active-matrix displays, and stretchable power sources. In addition, organic solar cells can be used as stretchable power sources. Electronic paper, flexible displays and touch screens, and robots with skin-like sensing capabilities are just a few examples of other cutting-edge products that have recently made their debut on the market.

In light of the recent emphasis placed on IoT devices and wear-able flexible sensors, there has been a renewed push for research into flexible electronics. Flexible electronics are electronics that can be bent or twisted, allowing them to be worn or installed on a variety of different things. For use in a variety of IoT applications, flexible ma-terials need to have a high level of mechanical resilience. This means they must be able to roll and bend multiple times without breaking. There are a variety of applications that can benefit from the utilization of flexible material, including RFID tags, wearable sensors, and flex-ible smart fitness watches. In addition, the elasticity and resilience of materials are crucial features required by electronic devices, as these devices require considerable and reversible deformation of the mate-rial. These versatile and flexible devices must also be able to store energy, operate with a minimal amount of power, and be compatible with other devices and applications for the IoT.

4.3.6 Printed Antennas

Printed antennas, also known as microstrip antennas, can be con-structed using your PCB. They have specific widths and lengths to match the impedance and frequency of the emitting circuit, and they are constructed using a geometric pattern that is laid out on the top copper plane. They are able to follow a variety of patterns, including the inverted F pattern, the straight trace pattern, the meandering in-verted F pattern, the circular pattern, and others.

One can think of a printed antenna as a simplified and scaled-down form of a traditional wire antenna. Printed antennas [18] are be-coming increasingly popular. It is possible for it to hold a wide range of frequencies and impedances depending on the dimensions of the copper planes and the microstrip traces that it contains.

Advantages include:

- Lower height profile;
- Bigger bandwidths;
- Stronger signals;
- Dependability;
- Low production costs for PCBs because they are printed di-rectly onto the circuit board.

Disadvantages include:

- Modifications to the design layout could result in a large financial setback because the entire PCB would need to be created again with the new design. This would require the expenditure of additional time and resources.

- The manufacturing of this antenna would require additional space on your PCB.

4.3.7 2-D and 3-D Antennas

Antennas can be classified based on their geometry and the number of dimensions that they occupy. Two common classifications are 2-D antennas and 3-D antennas.

Two-dimensional antennas are antennas that occupy a single plane and have a length and width, but no height. Examples of 2-D antennas include microstrip patch antennas, dipole antennas, and loop antennas. These antennas are typically used in applications where a low profile is desired, such as in mobile devices, wireless sensors, and RFID tags.

Three-dimensional antennas, on the other hand, occupy three dimensions and have a length, width, and height, as seen in Figure 4.7. The figure depicts an antenna held between two fingers.

Examples of 3-D antennas include helical antennas, Yagi antennas, and parabolic antennas. These antennas are typically used in

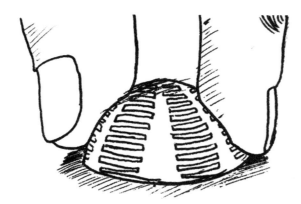

Figure 4.7 Three-dimensional printed antenna.

applications where a high gain and directional radiation pattern is required, such as in satellite communications, terrestrial broadcasting, and radar systems.

The main difference between 2-D and 3-D antennas is the way that they radiate energy. Two-dimensional antennas radiate energy primarily in the plane of the antenna, while 3-D antennas radiate energy in all directions. This means that 2-D antennas are typically less directional and have a lower gain than 3-D antennas, but they are also more compact and easier to integrate into devices. Designing a 2-D or 3-D antenna involves selecting the appropriate geometry and dimensions to achieve the desired performance characteristics. This involves selecting the appropriate frequency range, radiation pattern, gain, and impedance matching. Simulation tools such as finite element analysis (FEA) and MoM can be used to model the behavior of the antenna and optimize its performance.

In summary, 2-D antennas are antennas that occupy a single plane and are used in applications where a low profile is desired, while 3-D antennas occupy three dimensions and are used in applications where a high gain and directional radiation pattern is required. The choice of 2-D or 3-D antenna depends on the specific application requirements and design constraints.

Advantages of 2-D antennas are:

- *Low profile*: 2-D antennas are flat, making them ideal for use in devices with limited space.
- *Easy integration*: 2-D antennas can be easily integrated into PCBs and other electronics.
- *Low cost*: 2-D antennas can be manufactured using low-cost techniques such as PCB technology.

Disadvantages of 3-D antennas are:

- *Low gain*: 2-D antennas typically have lower gain than 3-D antennas.
- *Limited radiation pattern*: 2-D antennas typically have a limited radiation pattern, making them less suitable for directional applications.

Applications include:

- *Mobile devices*: 2-D antennas are commonly used in mobile devices such as smartphones, tablets, and laptops.
- *Wireless sensors*: 2-D antennas are used in wireless sensor networks for applications such as environmental monitoring, home automation, and industrial control.
- *RFID tags*: 2-D antennas are used in RFID tags for tracking and identification applications.

Advantages of 3-D antennas are:

- *High gain*: 3-D antennas typically have higher gain than 2-D antennas, making them suitable for applications that require long-range communication.
- *Directional radiation pattern*: 3-D antennas can be designed to have a highly directional radiation pattern, making them suitable for applications such as radar and satellite communications.

Disadvantages of 3-D antennas include:

- *Bulky*: 3-D antennas are typically bulkier than 2-D antennas, making them less suitable for use in devices with limited space.
- *Complex design*: 3-D antennas require more complex design and manufacturing processes than 2-D antennas.

Applications of 3-D antennas include:

- *Satellite communication*: 3-D antennas are used in satellite communication systems for long-range communication.
- *Terrestrial broadcasting*: 3-D antennas are used in terrestrial broadcasting systems for transmitting TV and radio signals.
- Radar systems: 3-D antennas are used in radar systems for detection and tracking of objects.

4.4 CONCLUSION

Antennas are essential components for wireless communication systems. There are various types of antennas available, each with their advantages and disadvantages.

Dipole antennas are the simplest and most common type of antenna, with a relatively broad radiation pattern. Patch antennas are low profile and suitable for integration into PCBs. Helical antennas have a high gain and can be designed for circular polarization. Yagi-Uda antennas are highly directional and suitable for long-range communication.

In addition to these common types of antennas, there are also specialized antennas such as microstrip antennas, slot antennas, and horn antennas, each with unique features and applications as discussed in this chapter.

The choice of antenna depends on the specific requirements of the wireless communication system, such as frequency range, gain, radiation pattern, and size constraints. A careful consideration of the trade-offs between different antenna types is essential to optimize the performance of the wireless communication system.

References

[1] Tyagi, D., S. Kumar, and R. Kumar, "Multifunctional Antenna Design for Internet of Things Applications," in *2021 7th International Conference on Advanced Computing and Communication Systems (ICACCS)*, Coimbatore, India, 2021, pp. 557–560, doi: 10.1109/ICACCS51430.2021.9441696.

[2] Balanis, C. A., "Fundamental Parameters and Definitions for Antennas," in *Modern Antenna Handbook*, Hoboken, NJ: Wiley, 2008, pp. 1–56, doi: 10.1002/9780470294154.ch1.

[3] Parthiban, P., "IoT Antennas for Industry 4.0—Design and Manufacturing with an Example," in *2020 IEEE International IoT, Electronics and Mechatronics Conference (IEMTRONICS)*, Vancouver, BC, Canada, 2020, pp. 1–5, doi: 10.1109/IEMTRONICS51293.2020.9216349.

[4] Zhang, Z., "Antenna Matching," in *Antenna Design for Mobile Devices*, IEEE, 2011, pp. 19–58, doi: 10.1002/9780470824481.ch2.

[5] de Lima, R. N., B. Huyart, E. Bergeault, and L. Jallet, "An Impedance Matching System Using a Line Loaded by Capacitors and Power Detectors," in *2000 30th European Microwave Conference*, Paris, France, 2000, pp. 1–3, doi: 10.1109/EUMA.2000.338670.

[6] Rahman, A., F. Olinger, and M. Howieson, "Rated Power Increment and VSWR Characteristics Improvement of Termination Resistor," in *2011 IEEE International Conference on Electro/Information Technology*, Mankato, MN, 2011, pp. 1–6, doi: 10.1109/EIT.2011.5978598.

[7] Khalily, M., M. K. A. Rahim, and A. A. Kishk, "Bandwidth Enhancement and Radiation Characteristics Improvement of Rectangular Dielectric Resonator Antenna," in *IEEE Antennas and Wireless Propagation Letters*, Vol. 10, 2011, pp. 393–395, doi: 10.1109/LAWP.2011.2144558.

[8] Yaghjian, A. D., and S. R. Best, "Impedance, Bandwidth, and Q of Antennas," in *IEEE Transactions on Antennas and Propagation*, Vol. 53, No. 4, April 2005, pp. 1298–1324, doi: 10.1109/TAP.2005.844443.

[9] Werner, D. H., "Exact Expressions for the Total Radiated Power, Radiation Resistance, and Directivity of an Arbitrary Size Uniform Current Elliptical Loop Antenna," in *IEEE Transactions on Antennas and Propagation*, Vol. 68, No. 9, Sept. 2020, pp. 6816–6820, doi: 10.1109/TAP.2020.2976528.

[10] Johnson, R. C., H. A. Ecker, and J. S. Hollis, "Determination of Far-Field Antenna Patterns from Near-Field Measurements," in *Proceedings of the IEEE*, Vol. 61, No. 12, Dec. 1973, pp. 1668–1694, doi: 10.1109/PROC.1973.9358.

[11] Warnick, K. F., and B. D. Jeffs, "Gain and Aperture Efficiency for a Reflector Antenna with an Array Feed," in *IEEE Antennas and Wireless Propagation Letters*, Vol. 5, 2006, pp. 499–502, doi: 10.1109/LAWP.2006.886308.

[12] Maheshwari, A., S. Behera, R. Thiyam, S. Maiti, and A. Mukherjee, "Near Field to Far Field Transformation by Asymptotic Evaluation of Aperture Radiation Field," in *2014 International Conference on Signal Propagation and Computer Technology (ICSPCT 2014)*, Ajmer, India, 2014, pp. 745–749, doi: 10.1109/ICSPCT.2014.6884958.

[13] Krishna, C. M., D. D. Prasad, C. R. Krishna, and M. K. V. Subbareddy, "Design and Analysis of Bow-Tie Antenna for Sub-6GHz Applications," in *2021 International Conference on Computer Communication and Informatics (ICCCI)*, Coimbatore, India, 2021, pp. 1–4, doi: 10.1109/ICCCI50826.2021.9402455.

[14] Kraus, J. D., "The Helical Antenna," in *Proceedings of the IRE*, Vol. 37, No. 3, March 1949, pp. 263–272, doi: 10.1109/JRPROC.1949.231279.

[15] Chen, Z. N., X. Qing, M. Sun, K. Gong, and W. Hong, "60-GHz Antennas on PCB," in the *8th European Conference on Antennas and Propagation (EuCAP 2014)*, The Hague, Netherlands, 2014, pp. 533–536, doi: 10.1109/EuCAP.2014.6901812.

[16] Chen, Z. N., M. J. Ammann, X. Qing, X. H. Wu, T. S. P. See, and A. Cai, "Planar Antennas," in *IEEE Microwave Magazine*, Vol. 7, No. 6, Dec. 2006, pp. 63–73, doi: 10.1109/MW-M.2006.250315.

[17] Chen, S. J., et al., "A Compact, Highly Efficient and Flexible Polymer Ultra-Wideband Antenna," in *IEEE Antennas and Wireless Propagation Letters*, Vol. 14, 2015, pp. 1207–1210, doi: 10.1109/LAWP.2015.2398424.

[18] Reig, C., and E. Ávila-Navarro, "Printed Antennas for Sensor Applications: A Review," in *IEEE Sensors Journal*, Vol. 14, No. 8, Aug. 2014, pp. 2406–2418, doi: 10.1109/JSEN.2013.2293516.

5

ANTENNA MEASUREMENT SYSTEMS

5.1 INTRODUCTION

Analog antenna testing is essential to understanding antenna theory. If the antennas being tested do not operate as expected, no amount of antenna theory will amount to much. Good antenna measurements do not just happen, they are a science unto themselves.

When testing or measuring antennas, what exactly is searched for? In essence, measuring any of the basic characteristics of an antenna, is the aim. The radiation pattern of an antenna, which includes antenna gain and efficiency, the impedance or VSWR, the bandwidth, and the polarization are the most often used and sought measurements [1].

The following sections cover the methods and tools used in antenna measurements.

5.2 EQUIPMENT NEEDED FOR ANTENNA MEASUREMENTS

The vector network analyzer (VNA) is the simplest basic tool for testing antennas [2]. A 1-port VNA is the simplest sort of VNA that can measure an antenna's impedance, which is comparable to measuring S11 and VSWR.

It is more challenging and expensive to measure an antenna's radiation pattern, gain, and efficiency. The antenna that needs to be measured will be referred to as the AUT. The following tools are necessary for antenna measurements:

- A reference antenna is an antenna with well-established properties (gain, pattern, etc.).
- An RF power transmitter is a device that transmits energy to the AUT.
- How much power is received by the reference antenna is determined by a receiver system.
- A positioning system rotates the test antenna with respect to the source antenna so that the radiation pattern can be measured as a function of angle.

Figure 5.1 displays a block diagram of the aforementioned equipment. A brief discussion of these elements is also discussed in coming sections. Of course, the reference antenna must emit effectively at the specified test frequency. To enable simultaneous measurements of horizontal and vertical polarization, reference antennas are frequently dual-polarized horn antennas.

The transmitting system should be able to emit a consistent, predetermined power level. Additionally, the output frequency must be selectable, stable, and customizable (stable means that the frequency from the transmitter is close to the expected frequency, and does not vary much with temperature). There will always be some energy outside of the desired frequency, but there shouldn't be much energy at harmonics, for example. The transmitter should have very little energy at all other frequencies.

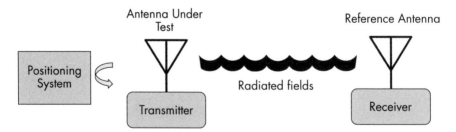

Figure 5.1 Antenna measurement setup.

The test antenna's orientation is controlled by the positioning system. The test antenna is rotated such that the source antenna illuminates it from every angle imaginable in order to measure the radiation pattern of the test antenna as a function of angle (usually in spherical coordinates). For this, the positioning system is employed. Figure 5.1 depicts the rotation of the AUT. It should be noted that there are various methods for carrying out this rotation; occasionally, both the reference and AUT antennas are turned.

Simply measuring the power received from the test antenna is all that is required by the receiving system. A straightforward power meter, which measures RF power and can be connected directly to the antenna terminals through a transmission line [3–5], can be used to accomplish this (such as a coaxial cable with N-type or SubMiniature Version A (SMA) connectors). The receiver is typically a 50-ohm system, but if another impedance is specified, it may be greater than 50 ohms. VNAs frequently take the place of transmit/receive systems. As seen in Figure 5.2, in an S21 measurement of a device under test (DUT), a frequency is transmitted from port 1 and the received power is recorded at port 2. As a result, a VNA is an effective tool for this purpose, however it is not the only one.

5.3   WHY MEASUREMENT LOCATION IS IMPORTANT

Where can accurate measurements of our antenna be taken? The reflections from the walls, ceilings, and floor would likely cause mea-

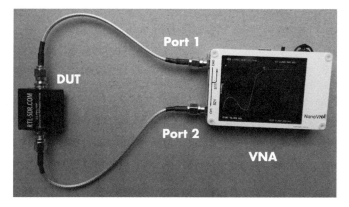

Figure 5.2 Measurement setup using VNA.

surements to be wrong if one attempted to perform this in a closed surrounding. Anywhere in space would be the ideal place to do antenna
measurements because there are no reflections there. RF-absorbing
foam can be used to absorb reflected radiation while isolating the antenna test equipment in an anechoic chamber.

5.4   RANGES IN FREE SPACE (ANECHOIC CHAMBERS)

Antenna measurement sites called free space ranges are created to
mimic measurements made in space. In other words, all unwanted
reflected waves from neighboring objects and the ground are as much
as possible subdued. Anechoic chambers [6, 7], elevated ranges, and
the compact range are the most widely used free space ranges.

5.4.1   Anechoic Rooms

Indoor antenna ranges are known as anechoic chambers. Special electromagnetic wave-absorbing material is used to line the floor, ceiling,
and walls. Because the test conditions may be precisely controlled,
indoor ranges are preferable to outdoor ranges. These chambers are
particularly interesting to observe because the material is frequently
shaped like jagged edges. The jagged triangular forms are made to
scatter whatever is reflected from them in random directions, and
whatever is added from all of the random reflections tends to accumu-
late incoherently and is so suppressed even more. Figure 5.3 displays
an anechoic chamber together with certain test instruments.

The disadvantage of using anechoic chambers is that they are
typically required to be rather large in size. Antennas almost always
need to be separated by a minimum number of wavelengths in order
for one to successfully imitate the conditions of a far-field environment. As a consequence of this, very large chambers are needed at
lower frequencies that have very large wavelengths; nevertheless, the
size of these chambers is typically limited by both practical and budgetary concerns. There are some defense contracting companies that
are known to have anechoic chambers the size of basketball courts in
order to measure the radar cross section of enormous aircraft or other
goods. However, this is a very rare occurrence. Typical dimensions
for anechoic chambers found in academic institutions are three to five
meters in length, width, and height. Because of their limited size and

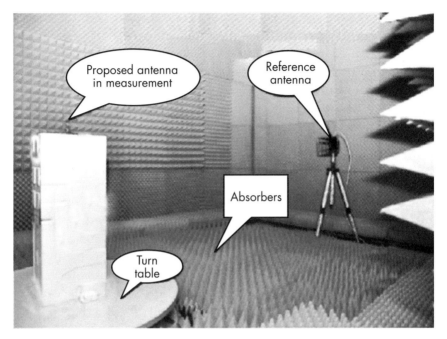

Figure 5.3   Anechoic chamber setup.

the fact that most RF absorption materials function most effectively at UHF and higher frequencies, anechoic chambers are utilized most frequently for frequencies that are greater than 300 MHz.

5.4.2   Heightened Ranges

Outdoor ranges are elevated ranges. In this design, both the source being tested and the antenna that it is attached to are elevated off the ground. These antennas can be installed virtually anywhere, including atop hills, towers, structures, or in any other appropriate setting [8]. When it would be hard to measure the antenna indoors, this method is typically used for extremely large antennas or at low frequencies (VHF and lower, 100 MHz). There is no guarantee that the source antenna, also known as the reference antenna, is located at a greater elevation than the test antenna. The LOS between the two antennas must be completely clear of any obstacles at all times. It is undesirable to have any other reflections, including those that originate from the Earth, such as the red ray. Following the selection of a source and a place for the test antenna in high ranges, the operators of the test will determine the surfaces that will cause significant reflections and then make an effort to decrease these reflections. In order to accomplish this goal, it is customarily necessary to make use of either RF-absorbing material or other materials that can redirect the rays away from the test antenna.

5.4.3 Smaller Ranges

It is imperative that the source antenna be positioned inside the far field of the test antenna. It is recommended that the wave that the test antenna receives be a plane wave for the highest level of accuracy. Because antennas generate spherical waves, the receiving antenna needs to be located at a sufficient distance from the source antenna for the wave it emits to be approximately equivalent to a plane wave.

On the other hand, there is typically not enough separation between indoor chambers for this to be possible. One possible solution to this problem is to make use of a portable cooking range. In this method, a source antenna is aimed in the direction of a reflector, and the design of the reflector is intended to cause the spherical wave to be reflected in a manner that is roughly planar. This works on a principle that is quite comparable to that of a dish antenna, which is to say that it is quite analogous to it.

It is common practice to require that the length of the parabolic reflector be significantly greater than that of the test antenna. To prevent the source antenna from obstructing the rays that are being reflected, the position of the source antenna is kept at a distance from the reflector. In addition, care must be taken to prevent any mutual coupling, also known as direct radiation, from occurring between the source antennas and the test antennas.

For instance, suppose the antenna that is being tested is illuminated by pointing a plane wave that originates from the source antenna in a certain direction. It is essential to have a solid understanding of the polarization and gain of the source antenna (for the fields transmitted toward the test antenna).

As a consequence of reciprocity, the transmission and reception modes of the test antenna both produce an identical radiation pattern from the antenna. Therefore, radiation pattern of the test antenna can be evaluated in both the receive and the transmit modes. Next, the receiving situation will be discussed that is associated with the test antenna. The rotation of the test antenna is accomplished with the help of the positioning system. At each step, the amount of power that was received is recorded. The magnitude of the test antenna's radiation pattern can then be determined as a result of this measurement. Measurements of polarization and phase will be taken into consideration later. For the radiation pattern, the spherical coordinate system is the most effective choice for the coordinates.

5.5 DIRECTIVITY AND EFFECTIVENESS

Regardless of the gain, the directivity may be determined from the recorded radiation pattern. This is typically accomplished by approximating the integral as a finite sum, which is a rather straightforward method. Mathematically, the antenna parameter equations [9] in the following paragraphs describe the relationship between the antenna's physical dimensions and its electrical performance. An antenna's efficiency is just the ratio of peak gain to peak directivity as given in:

$$\varepsilon = \frac{G}{D} \tag{5.1}$$

So, once we have measured the radiation pattern and the gain, we may deduce the efficiency from these measurements. We require both the magnitude of the power transmitted or received in order to fully specify the radiation pattern of an antenna [10]. To measure all of the antenna's components (polarizations), these measurements must be given in two orthogonal orientations.

Consider the following if an antenna were to transmit at frequency f and the fields traveling in the $+y$ direction were seen at a certain location as in:

$$E = \hat{x}Ae^{jD}e^{j2\pi ft} + \hat{z}Be^{jF}e^{j2\pi ft} \tag{5.2}$$

In the distant field region, the E-fields are orthogonal to the direction of movement. The x-component of the E-field has a magnitude of A, while the z-component has a magnitude of B. The x-phase's component is D, and the z-phase's component is F (relative to the oscillation at frequency f). If $D = F$, the polarization is linear and the components are in phase. The E-field is circularly polarized if D and F are 90 degrees apart and their amplitudes are equal.

The phase must be measured relative to a fixed reference because it is a relative quantity. The approach depicted in Figure 5.4 is the simplest for measuring phase. In this technique, a different antenna is utilized to receive the fields while the test antenna serves as the source antenna. The observation point must be close enough to the test antenna for this method to be effective so that a phase measurement box may be connected to the source waveform feeding the test antenna. This box compares the positions of the received signals' peaks and valleys and derives the relative phase from that data. The operation is repeated after the receiving antenna has been repositioned.

In order for the probe antenna to be able to pick up one component of the received field, it needs have good polarization purity. It may simply be turned, or another probe antenna might be used to obtain the second orthogonal component. The reference (source) waveform cannot be fed directly into the phase measurement circuit when the test antennas are very far apart (this occurs at low frequencies and large outdoor ranges where many wavelengths become very far apart). In this case, a standard antenna with known phase characteristics is used to transmit a wave that is used to compare with the signal received from the test antenna.

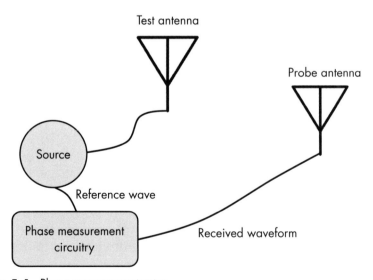

Figure 5.4 Phase measurement setup.

The polarization of an antenna is crucial to its radiating pattern. It is to be noted that the polarization changes are based on the antenna's radiating direction. For instance, a circularly polarized antenna may have linear polarization away from the antenna's primary beam and be approximately circular only across a short beamwidth (this is often the case for circularly polarized patch antennas). The test antenna is used as the source for the measurement. The receiving antenna will thus be a linearly polarized antenna (usually a half-wave dipole antenna). In order to record the received power as a function of the receive antenna's angle, the linearly polarized receiving antenna will be rotated. Only the polarization of the test antenna for the direction in which the power is received is covered by the received information. The test antenna must be rotated in order to ascertain the polarization for each direction of interest in order to provide a thorough description of the polarization of the test antenna. Figure 5.5 depicts the essential configuration for polarization measurements.

The power of the receiving antenna is first measured in a fixed position (orientation), after which it is rotated around the x-axis as depicted in Figure 5.5, and the power is measured once more. The linearly polarized receiving antenna is rotated entirely in this manner.

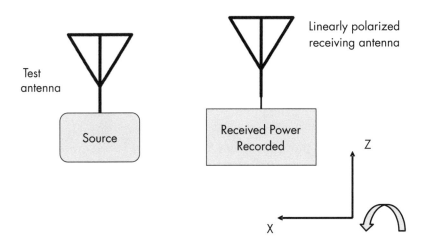

Receiving antenna rotation about X axis

Figure 5.5 Polarization measurement setup.

Much can be inferred about the polarization of the test antenna from this data. Let's examine a few cases. Assume that the receiving antenna and the test antenna are both vertically linearly polarized, and that at the rotation angle zero, the polarization of both antennas is matched. It is important to note that the outcome is periodic because the received power is the same whether the receive antenna is turned 180 degrees or not.

Now imagine that the test antenna was initially not polarization matched to the receiving antenna and was horizontally polarized, once more linearly polarized. In this instance, it is observed that the measurements have the same shape but that the received power peaks occur at various angles. The test antenna is linearly polarized, and we can determine the angle of the linear polarization by looking at the angle at which the received power peaks.

Now imagine that the RHCP wave was being transmitted by the test antenna. If the test antenna were subjected to the same measurement as before, that normalizes (makes the peak output power equal to one for simplicity) the output power mirrors.

The received power for a rotated linearly polarized antenna is constant for a circularly polarized wave [11] because it has equal amplitude components in two orthogonal directions (incidentally, this is a feature of circular polarization that makes it appealing because you don't have to worry about getting the orientation right). Also take note that whether the test antenna is in the left hand (LHCP) or right hand (RHCP) orientation, the received power is the same (RHCP). As a result, while using this method it is possible to identify the kind of polarization, but the sense of rotation for the polarization cannot be identified.

Suppose that the test antenna is elliptically polarized, has a 45-degree tilt, and an axial ratio of 3 dB. In the coordinate system shown in Figure 5.5, the following might be used to describe the E-field for this antenna:

$$E = \hat{y}\left(1 - \frac{j}{2}\right) + \hat{z}\left(1 + \frac{j}{2}\right) \tag{5.3}$$

The first way to determine the tilt angle of an elliptically polarized wave is to look at where the peak of the received power is, in this case, at 45 degrees. The axial ratio (1/0.5 = 2 = 3 dB) is calculated as the ratio of the maximum output (in this case, 1.0 when the angle is 45 degrees) to the minimum output (in this case, 0.5 when the angle is 135 degrees). As a result, we can immediately identify the kind of polarization for this direction of the test antenna's radiation pattern by merely looking at the plots and being familiar with the different types of polarization. Note once more that we are unsure of the E-rotational field's direction (left or right).

Finally, we want to establish the AUT's sense of polarization. Assume the test antenna is RHCP. Whether the outcome is RHCP or LHCP is unknown from this. Using an antenna that is known to be RHCP as the receiving antenna is an easy way to detect the sense of rotation in this situation. After recording the outcome, the experiment is repeated using an LHCP receive antenna. If the test antenna is RHCP, the output for the first instance should be significantly bigger than the output for the second case. The sense of rotation for the polarization can be found by choosing the polarization sense based on the output power that is greater. This is challenging since it calls for two antennas that are tightly spaced between RHCP and LHCP at the desired frequency, which isn't always simple.

A combination of phase measurements for two orthogonal radiation pattern directions and comparison of the results with the received power might also be used to calculate the polarization.

5.6 IMPEDANCE MEASUREMENTS

Impedance is a crucial quality of antennas, and it is one of the characteristics that distinguishes them. If the impedance of an antenna is not close to the impedance of the transmission line, which is normally 50 ohms, then the antenna will either receive very little power or will broadcast very little power (if the antenna is in transmit mode) if used in the receive mode. In the absence of the appropriate impedance, the antenna won't work properly (or an impedance matching network).

Antenna impedance and radiation pattern can be altered by the objects that are in its immediate vicinity. For example, an antenna can be constructed such that it is incorporated into a variety of other metallic components or so that it rests on top of a metal surface (such as an antenna found on an airplane), as in a mobile phone antenna. For this reason, the impedance ought to be measured in a setting that will most nearly approximate where it will operate, so as to get the highest level of precision possible. The impedance of an antenna in free space is referred to as its self-impedance or its stand-alone impedance. This impedance is measured when there are no adjacent objects that could impact the antenna's radiation pattern. In certain contexts, the term *driving point impedance* is used to refer to the input impedance that has been measured in a particular environment.

Taking impedance measurements does not present too much of a challenge so long as the appropriate instruments are used. A VNA is an important piece of equipment to have in a scenario like this one. One is to determine the input impedance as a function of frequency with the help of this instrument. There is also the possibility that it will represent S11 (return loss), which is a frequency-dependent function of the impedance of the antenna, in addition to the VSWR.

There are various VNAs available that cost less money and do not include an integrated display; one example is the VNA that Copper Mountain Technology offers. Coaxial cables with N-type to SMA connectors are used for the purpose of linking the VNA to the test antennas. A wideband dipole antenna as well as a horn antenna are both subjected to measurement. Figure 5.6 illustrates a Rohde & Schwarz VNA.

Let's pretend for the moment that we are interested in determining the impedance between 400 and 500 MHz. The very first thing that we need to do is check to see if our VNA is intended to function inside this frequency range. Network analyzers are able to function over specific frequency ranges, the lowest of which is approximately

Figure 5.6 Rohde & Schwarz VNA.

30 kHz, and the highest of which is in the gigahertz range (110 GHz or so, and these tend to cost in the hundreds of thousands of U.S. dollars). If we are confident that our network analyzer is functioning appropriately, then we can proceed.

After that, the VNA needs to have its calibration done. This is much simpler to accomplish than it may at first appear. In order to calibrate out the influence of the coaxial cables, which act as transmission lines and connect the VNA to the antenna, we will take the coaxial cables from the probes and use them to connect the probes to the antenna. This will allow us to calibrate out the influence of the coaxial cables. For this purpose, your VNA will typically come with a "cal kit" that consists of a matched load with a resistance of 50 ohms, an open circuit load, and a short circuit load. We seek for a calibration button on our VNA and go through the menus until we find it. Once we do, we click the button and proceed to follow the instructions. In order for the VNA to know what to anticipate with your wires, it will ask you to attach the supplied loads to the ends of your cables while it is recording data. This is done so that it can determine what to anticipate with your wires. You are finished once you have applied the three loads when told to do so. You don't even need to be an expert; all you have to do is follow the instructions that the VNA gives you, and it will do all the computations for you. In addition, there are automatic e-cal kits available that utilize switches to carry out the open/short/thru measurements that you want.

At this point, the VNA should be connected to the test antenna. Choose the frequency range that interests you most and enter it into the VNA. Because there are so many buttons, if you are unsure how to use it, you should just experiment with it until you figure it out; there is no real risk involved.

If you ask for the output in the form of an S-parameter, the return loss will be measured (S11). In this instance, the VNA will send a relatively low amount of power to your antenna, and then measure the amount of power that is reflected back to the VNA. The impedance of the antenna as well as the impedance of the VNA, which is typically set to 50 ohms, will have an effect on the S-parameter. The S-parameter measures the magnitude of the reflection coefficient. Because of this, this measurement is typically used to determine how close the antenna impedance is to the value of 50 ohms.

Another frequent output is to use a Smith chart to determine the impedance of the circuit. In its most fundamental form, a Smith chart is a straightforward and pictorial representation of the input impedance (or reflection coefficient) that displays amplitude and phase information simultaneously. Due to the fact that the center of the Smith chart represented a reflection coefficient of zero, the antenna has been appropriately matched to the VNA. The outer border of the Smith chart reveals a reflection coefficient with a magnitude of 1, which indicates that the antenna and VNA are not matched very well. This indicates that all of the power is reflected. The magnitude of the reflection coefficient (which must be kept to a minimum in order for an antenna to successfully receive or broadcast signals) varies depending on how far away you are from the center of the Smith chart. After being measured across a range of frequencies, the reflection coefficient is represented on a Smith chart. This chart displays the data in Figure 5.7.

At the point on the Smith chart where there is a zero reflection coefficient, the impedance of the antenna would be properly matched to the impedance of the generator or receiver. The measurement is depicted by the curved red line in the image. The impedance of the antenna is illustrated in a frequency spectrum ranging from 2.7 GHz to

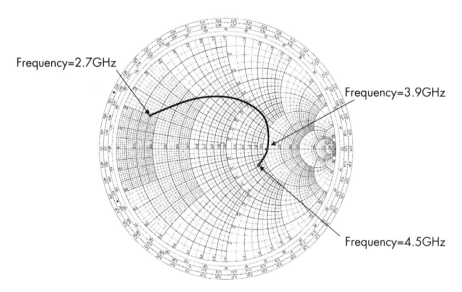

Figure 5.7 Smith chart.

4.5 GHz. Each point on the line is a representation of the impedance at a particular frequency. Inductive impedances, which have a positive reactance, are denoted on the Smith chart by points that are located above the equator (imaginary part). The locations on the Smith chart that are located below the equator illustrate capacitive impedances, which have a reactance that is in the opposite direction from zero (for instance, the impedance would be something like $Z = R - jX$).

The impedance at $f = 4.5$ GHz, which is represented by the dot that can be found underneath the equator in Figure 5.7, will help to better explain the figure. Since the dot is located exactly one-quarter of the way between the origin and the outer perimeter, the reflection coefficient, which measures the distance from the origin, can be calculated to have a magnitude of approximately 0.25.

As the frequency decreases, there is a corresponding change in the impedance. On the impedance test, the frequency $f = 3.9$ GHz corresponds to the position of another dot. At this location, the antenna has reached its resonance frequency, indicating that the impedance is at its absolute maximum. The process of scanning the frequency down until $f = 2.7$ GHz creates the locus of points, which illustrates the antenna impedance over the frequency range. The curve represents this locus of points. Given that it is located closer to the perimeter of the Smith chart than the center at $f = 2.7$ GHz, the impedance is inductive, and the reflection coefficient is roughly 0.65; the frequency is higher than the center.

To summarize, the Smith chart is a useful tool that allows one to rapidly and clearly visualize impedance across a range of frequencies. Last but not least, VSWR is widely employed as a measurement tool for determining the magnitude of an impedance mismatch. The impedance's phase cannot be established because the value of the VSWR is dependent on the magnitude of the reflection coefficient. The VSWR provides a quick approach for determining the amount of power that is reflected by an antenna. As a consequence of this, VSWR is cited quite frequently in antenna data sheets, such as "VSWR: 3:1 from 100–200 MHz." According to the VSWR calculation, this means that less than half of the power is reflected from the antenna over the stated frequency range. In other words, the antenna is not reflecting the full amount of power. The S11, Smith chart, and VSWR measurements all make use of the same type of measurement; however, the format can differ according to what the user prefers.

5.7   ISOLATION MEASUREMENTS

In Section 5.6, VNA measurements were done with a single port. Often VNAs are capable of measuring 2 ports (sometimes 4 ports, and there are some that go as high as 24 ports!). If you connect an antenna to each port (for example, as shown in Figure 5.2), then S11 is the impedance of the antenna on port 1, and S22 is the impedance of the antenna on port 2. The VNA also measures the coupled power between the antennas, which is S12 or S21 (for antennas, these values will be identical due to reciprocity). S12/S21 is commonly referred to as isolation. Isolation [12] is a measure of how strongly the antennas couple together.

For example, if you have two Wi-Fi antennas in your product, you want them to have a high isolation (it's debatable on the exact value required, but below −15 dB is a good number). High isolation means the antennas operate independently. Low isolation (such as S12 larger than −8 dB) means the antennas couple together strongly, which degrades their antenna efficiency. Low isolation also means the antennas correlate very strongly, which ultimately defeats the purpose of having multiple antennas.

In summary, impedance is measured with a VNA, and can be displayed in various ways including the Smith chart, VSWR, and S11. We care about the impedance of an antenna so that we can properly transfer/receive power to/from the antenna from/to the transmitter/receiver. VNAs enable us to measure an antenna's impedance (the V in VNA means vector, so that we can measure magnitude and phase) as well as the isolation between different antennas.

5.8   SCALE MODEL MEASUREMENTS

When measuring the performance of an antenna, it is important to do it in a setting that is as close as possible to the environment in which it will be used in order to get the most accurate picture possible of how well it will function in the real world. However, there are some circumstances in which it is required to make precise measurements [13] that are not possible to make in the real world. Consider the scenario in which an antenna is going to be installed on an airplane, and we are interested in the radiation pattern, the impedance of the antenna, and other such things. Or, we are interested in the coupling between two antennas that are both connected to an arbitrary plane. In the latter case, we would have to consider that the process of checking the antennas after they have been mounted on an airplane is not only complex but also very challenging and expensive (particularly the radiation pattern). In addition, measurements are frequently requested prior to finalizing antenna sites. Because of this, we would like to try to locate an antenna in a few different locations to see which one has the field of view (FOV, the directions that the aircraft is not hidden from with respect to the antenna) and gain that we require.

In this scenario, one technique that we might take is to perform measurements using a scale model. A scaled model, which is typically a substantially smaller physical representation of the actual structure, is used in this method to stand in for the platform on which the antenna operates. This model is called a scaled-down model. We have to ask: Is this really the case? Can a smaller model produce reliable measurements?

The answer to this question is yes, provided that we are aware of what we are doing. The most fundamental mystery of the universe is that every signal may be interpreted as a combination of frequencies. This is one of the many ways in which signals can be represented. In addition, the theory of antennas is encompassed under Maxwell's equations, which addresses all aspects of electromagnetics. If you were traveling along a monochromatic (wave of a single frequency) electric field, your behavior would only be affected by the incident wavelength size if you came across an obstruction. In other words, the wave that is produced by a perfectly conducting circular plate that has a diameter equal to three wavelengths will be the same regardless of the frequency that is being used.

So, suppose we are interested in finding out more information on the characteristics of a monopole antenna operating at 300 MHz (wavelength is 1 meter). Imagine that this antenna is attached to a plane that is 30 meters, or 30 wavelengths, in length. Thought should be given to the construction of a scale replica of our airplane measuring three meters in length that can be accommodated within the anechoic chamber. In order to accurately replicate the electromagnetic waves that go through a real airplane at a frequency of 300 MHz, the scale model needs to be 30 wavelengths long. If we work at a frequency of 3,000 MHz, the model is now 30 wavelengths in length (3 GHz, where the wavelength is 0.1 meters). If we scale the monopole antenna by the same factor, then the results of measurements taken at 3 gigahertz on the model will theoretically be equivalent to those taken at 300 megahertz on the actual airplane.

The accuracy of the model is directly proportional to the accuracy of the results. Measurements of antennas can be carried out in a dependable and common fashion with the help of this method, which is utilized in the aerospace and defense industries.

Table 5.1 is offered for your use in better comprehending correct scaling. It may be deduced from the table that the model has been reduced in size by a factor of n. For example, the model of the airplane that was reduced in size and utilized in the previous statement has a scale factor of $n = 10$. For example, if $n = 100$, a scaled model with a size of $L/n = 1$ meter would be used to show a real-world object with a length of $L = 100$ meters; this would be possible because of the model's smaller size.

Table 5.1 is the result of carrying out the mathematical calculations in Maxwell's equations. The good news is that scaling is not necessary at all for certain characteristics, such as permittivity and permeability in particular (if you are modeling dielectrics or magnetic materials). An additional advantage is that the antenna's impedance as well as its gain do not need to be scaled. In the scaled model, it is necessary to multiply certain factors by n, such as the conductivity factor. This can be understood by keeping in mind that the resistance

Table 5.1
Relationships Between Scaled
Measurements and Their Quantities

Quantity	Scaled Relationships
Length (L)	L/n
Frequency (f)	$f * n$
Permittivity (ε)	ε
Permeability (μ)	μ
Conductivity (σ)	$\sigma * n$
Impedance (Z)	Z
Gain (G)	G
Radar cross section (A)	$A * n^2$
Capacitance (C)	C/n
Inductance (L)	L/n

of the scaled-down model should remain the same and that resistance has an inverse relationship with conductivity multiplied by length and then divided by the item's cross-sectional area. We require the conductivity on the scaled model to increase by a factor of n if the length drops by a factor of n and the cross-sectional area decreases by a factor of n^2. This is because we want the resistance to remain constant. Airplanes that are used in the actual world frequently have shells that have a high conductivity (metals). As a result, extreme vigilance is required to guarantee that the model's conductivity is as high as is practically possible. This can be accomplished for the scale model by using metals of a high quality that have been polished.

5.9   SPECIFIC ABSORPTION RATE

The specific absorption rate, often known as SAR, is the rate at which RF energy emitted from a device is absorbed by human tissue. SAR [14] is a function of the tissue's mass density (measured in kg/cubic-meter), electrical conductivity (measured in siemens/meter), and the induced E-field from the radiated energy (measured in volts/meter). In order to calculate the SAR, a volume (often a 1-gram or 10-gram region) is integrated or averaged as shown:

$$SAR = \int_{sample} \frac{\sigma(r)|E(r)|^2}{\rho(r)} \tag{5.4}$$

The SAR is expressed as a ratio of watts per kilogram to milli-watts per gram. The SAR limit in the United States for mobile phones is 1.6 W/kg, which is averaged throughout 1 gram of tissue. In Europe, the maximum SAR allowed is 2.0 watts per kilogram, averaged across 10 grams of tissue.

5.10   SAR MEASUREMENT

SAR is measured using standardized forms that have been modeled taking the human head shape into consideration. The mobile phone, which will act as the DUT, will be in the vicinity head shape mod-el and will transmit at the highest possible strength throughout the whole test. In order to carry out the averaging that is outlined in (5.4), the probe is moved through the head region. This set up can be implemented using a robot arm.

In order to faithfully reproduce the conductivity and density of human tissue, the model is stuffed with a liquid that has a similar appearance to human tissue. Since the properties of the fluids change depending on the frequency, a frequency-specific standard fluid for a measurement at 1,800 MHz will be different from the fluid that should be used for a SAR measurement at 900 MHz. In spite of the fact that the measurements are nearly equivalent in terms of geometry, the results can nevertheless differ substantially due to the chaotic features of the near field. The maximum SAR values that were measured from the left and right sides of the head for each frequency that a mobile phone operates in are the values that are reported as the phone's SAR rating.

When designing an antenna, it is necessary to bring down the SAR value since, if it is too high the antenna will need to be altered [15]. When the SAR is too high, the transmit power is often dropped, which directly results in a reduced SAR. Despite the fact that the minimum transmit power needs for mobile devices are quite low, the SAR cannot be eliminated completely. Because of this, the placement of the antenna is quite important. In order to keep the radiating component of mobile phones as far away from the human brain as is practically possible, antennas are typically positioned near the bottom of the device.

Adjustments to the impedance matching and the use of parasitic resonators, both of which will cause a disruption in the radiation pattern of the antenna, are two more options for lowering the SAR.

5.11 CONCLUSION

Antenna measurements play a crucial role in the characterization and optimization of antenna performance. Accurate measurements are essential for evaluating key parameters such as gain, radiation pattern, polarization, and impedance matching. Various techniques, including near-field and far-field measurements, have been employed to capture these characteristics. Advanced measurement setups, such as anechoic chambers and network analyzers, have facilitated precise and reliable antenna measurements. Additionally, advancements in measurement technologies, such as MIMO systems and beamforming, have further expanded the scope of antenna measurements. These measurements are vital for the design and development of antennas in various applications, including wireless communication systems, radar systems, satellite communication, and IoT devices.

References

[1] Jang, G., S. Park, H. Keum, S. Choi, and G. Kim, "Antenna Measurement System for 5G Application," in 2018 *International Symposium on Antennas and Propagation (ISAP)*, Busan, Korea (South), 2018, pp. 1–2.

[2] Kiraz, E., F. A. Hatık, and S. Günel, "Standalone Antenna Measurement System," in *2022 International Conference on Applied Electronics (AE)*, Pilsen, Czech Republic, 2022, pp. 1–5, doi: 10.1109/AE54730.2022.9920022.

[3] "IEEE Recommended Practice for Antenna Measurements," in *IEEE Std 149-2021 (Revision of IEEE Std 149-1977)*, Feb. 18, 2022, pp. 1–207, doi: 10.1109/IEEESTD.2022.9714428.

[4] Monebhurrun, V., "IEEE Standard 149-2021: IEEE Recommended Practice for Antenna Measurements [Stand on Standards]," in *IEEE Antennas and Propagation Magazine*, Vol. 64, No. 3, June 2022, pp. 143–143, doi: 10.1109/MAP.2022.3162793.

[5] "IEEE Recommended Practice for Near-Field Antenna Measurements," in *IEEE Std 1720-2012*, Dec. 5, 2012, pp. 1–102, doi: 10.1109/IEEESTD.2012.6375745.

[6] Rodriguez, V., "Comparing Predicted Performance of Anechoic Chambers to Free Space VSWR Measurements," in *2017 Antenna Measurement Techniques Association Symposium (AMTA)*, Atlanta, GA, 2017, pp. 1–6, doi: 10.23919/AMTAP.2017.8123694.

[7] Sirles, C. W., J. C. Mantovani, A. R. Howland, and B. J. Hart, "Anechoic Chamber Performance Characterization Using Spherical Near-Field Imaging Techniques," in *2009 3rd European Conference on Antennas and Propagation*, Berlin, Germany, 2009, pp. 1734–1738.

[8] Gupta, I. J., and W. D. Burnside, "Compact Range Measurement Systems for Electrically Small Test Zones," in *IEEE Transactions on Antennas and Propagation*, Vol. 39, No. 5, May 1991, pp. 632–638, doi: 10.1109/8.81491.

[9] Kraus, J. D., R. J. Marhefka, and A. S. Khan, *Antennas and Wave Propagation*, 5th edition, McGraw Hill Education, 2017.

[10] Sapuan, S. Z., A. Nasimuddin, A. Kazemipour, M. Z. Mohd Jenu, and M. F. Hassan, "A Fast and Accurate Analytical Method Based on Input Impedance for Antenna Directivity [Antenna Applications Corner]," in *IEEE Antennas and Propagation Magazine,* Vol. 60, No. 4, Aug. 2018, pp. 94–99, doi: 10.1109/MAP.2018.2839961.

[11] Cakaj, S., and K. Malaric, "Isolation Measurement Between Uplink and Downlink Antennas at Low Earth Orbiting Satellite Ground Station," in *2007 19th International Conference on Applied Electromagnetics*

and Communications, Dubrovnik, Croatia, 2007, pp. 1–4, doi: 10.1109/ICECOM.2007.4544490.

[12] Aloi, D. N., and M. Alsliety, "A Methodology to Determine the Isolation Requirements Between Collocated Cellular and GPS Antennas in Telematics," in *IEEE Antennas and Wireless Propagation Letters,* Vol. 6, 2007, pp. 1–4, doi: 10.1109/LAWP.2006.890744.

[13] Onishi, T., K. Kiminami, and T. Iyama, "Novel Specific Absorption Rate Measurement Techniques," in *2008 Asia-Pacific Symposium on Electromagnetic Compatibility and 19th International Zurich Symposium on Electromagnetic Compatibility,* Singapore, 2008, pp. 120–123, doi: 10.1109/APEMC.2008.4559826.

[14] Zhang, Y., and V. Monebhurrun, "Numerical Investigation of the System Validation Setup Adopted for Specific Absorption Rate Measurement," in *2023 IEEE Radio and Antenna Days of the Indian Ocean (RADIO),* Balaclava, Mauritius, 2023, pp. 1–2, doi: 10.1109/RADIO58424.2023.10146072.

[15] Okano, Y., and H. Shimoji, "Comparison Measurement for Specific Absorption Rate with Physically Different Procedure," in *IEEE Transactions on Instrumentation and Measurement,* Vol. 61, No. 2, Feb. 2001, pp. 439–446, doi: 10.1109/TIM.2010.2045939.

6

MINIATURIZATION TECHNOLOGIES

6.1 INTRODUCTION

For more than 50 years, there has been a lot of interest in the possibility of reducing the physical size of an antenna without noticeably degrading its performance. Numerous antenna-shrinking methods have been put forth over the years, including structural adjustments, lumped component loading, the use of materials with high permittivity and permeability, and more recent uses of metamaterials. The demand for compact antennas has grown along with consumer demand for cellular communications, digital media streaming, and social connectedness. The need for integrated antennas, which are often much smaller than a wavelength in size and can work with systems on low power budgets, has increased due to the desire to connect all electronic systems, even the smallest of items, into an IoT. Additional motivation for antenna and related electronics size reduction comes from wireless sensor networks, intelligent vehicles, and dwindling military systems like mini-UAVs. Since Heinrich Hertz proved that radio waves exist, small antennas have been used. However, historically, the creation of complex antenna systems with suitable operating properties has depended on having an aperture with enough size and volume. The idea behind antenna shrinking is to scale down these

tested systems while reasonably preserving some of the original antenna's performance. The task of miniaturizing a range of traditional
antennas and classes of antennas, such as loops, dipoles, slots, patches, leaky-wave, and various integrated antennas, is challenging and
intimidating, yet antenna experts have devised a remarkable number
of inventive strategies. Making a metamaterial antenna or otherwise
modifying the resonant characteristics of antennas using the broad
principles of metamaterials is one strategy that has attracted a lot of
interest recently (producing metamaterial-inspired antennas). Purely
using simulation software or analytical methods, many miniaturization strategies have been studied. Most, if not all, of the typical performance characteristics, such as impedance, bandwidth, efficiency,
or gain, are typically reduced with small antennas. A miniaturized
antenna, on the other hand, is one from a well-defined category that
has had its size decreased while maintaining the accuracy of at least
one performance parameter. This is accomplished by modifying the
structure's geometry, including extra parts, or changing the properties
of the materials. Although the performance feature need not be as
good as the original antenna's, it must be superior to a similarly sized
unmodified antenna.

6.2   PRACTICALITIES IN MINIATURIZATION

Not every antenna is a resonant antenna, and in order to reach smaller
sizes, the structure of the antenna is frequently altered. Quantification,
on the other hand, becomes increasingly difficult as miniaturization
occurs. An antenna consists of three dimensions, and the size of the
antenna can be decreased in any combination of these dimensions.
For instance, the size of a patch antenna can be reduced by reducing
the area that it occupies, and the size of a leaky-wave antenna can be
reduced by reducing the length that it stretches across. It is common
practice to shrink one dimension while simultaneously growing another. Although folding a patch antenna results in an increase in its
thickness, doing so may result in a reduction in the antenna's total
size.

As a consequence of this, the following methodology is applied
to determine size reduction wherever it is feasible to do so:

- The length reduction is the difference between the linear dimension of the original antenna and a particular linear dimension of the smaller antenna.

- The smaller antenna has a shorter overall linear dimension. The term *area reduction* refers to the process by which a small antenna's overall surface area is reduced in comparison to the size of the antenna's initial surface area. The term *volume reduction* refers to the percentage of the larger antenna's volume that the newer, smaller antenna occupies in contrast. With the help of this strategy, scenarios in which reducing the value of one size parameter could be advantageous, despite the fact that doing so would necessitate increasing the value of another can be considered. For instance, when a small antenna must be integrated onto a circuit board, it may be advantageous to reduce the area by 50%, even while the thickness is doubled, and the volume remains unchanged. This can be the case even though the volume does not change. Every factor in the miniaturization process is expressed as a ratio, such as 3.2/1 or 1/7.5. If the ratio is less than one, the geometrical quantity that is being given (whether it be length, area, or volume) has decreased. If the ratio is greater than one, then the magnitude of the ratio has increased. When an antenna needs to be connected to a massive neighboring structure in order for it to function properly, such as when an integrated antenna needs a ground plane that is larger than the major radiator, this is one situation that can be particularly challenging. In this scenario, an antenna must be connected to the neighboring structure. It is possible that reducing the size of the main radiator while maintaining the size of the neighboring structure might be beneficial; nevertheless, it is undeniably more beneficial to reduce the size of both the main radiator and the neighboring structure. A shorter length, a smaller surface area, or a smaller volume are all examples of what is meant by the term *primary reduction*.

- The term *secondary reduction* refers to shrinking the size of a substructure of an antenna system while keeping the dimensions of adjacent components the same (or nearly unchanged). An example of this would be a patch antenna, in which the ground plane and patch radiator both have the same dimensions, but the patch radiator has a lower surface area. Expanding the current architecture of an antenna in order to minimize the size of the main radiator is an example of another situation that may occur on occasion. One example of this would be the installation of a dipole antenna inside of a substantial ball made of man-made material. If the ball were larger than the original dipole and the resonance frequency stayed the same, it is difficult to imagine a scenario in which the addition of the ball would be useful; still, conducting research on the topic would be interesting. One possible name for this phenomenon is secondary decline.

6.2.1   Performance Limitations for Tiny Antennas

Wheeler [1] and Chu [2, 3] demonstrated a direct correlation between antenna bandwidth and antenna size for a lossless antenna by establishing theoretical lower bounds on the Q of tiny antennas in the 1940s. This also suggests that bandwidth and effectiveness are mutually exclusive (or realized gain). The trade-off between bandwidth and efficiency must be maintained as antennas get smaller; more bandwidth necessitates poorer efficiency, and vice versa. For antenna designers, this is a major practical concern.

6.2.2   Miniaturization Measurement

The resonance frequency, f_0, which is typically correlated with the size of the antenna construction, has a small range of operational frequencies that are centered around it. Some people describe miniaturization as decreasing the resonance frequency to a value of f_1 without altering the antenna's size. One way to achieve this is by adding inductance. The definition of a miniaturization factor [4] m is then:

$$m = \frac{f_1}{f_2} \tag{6.1}$$

6.3   METAMATERIALS-BASED MINIATURIZATION

Synthetic materials known as metamaterials are distinguished by the unique electromagnetic behavior they exhibit in the vicinity of resonance. Inserting arrays of metallic or dielectric inclusions into a host medium at a lattice spacing that is much smaller than the operational wavelength is the procedure that is used to create metamaterials. In the past 10 years, there has been an explosion of interest in the amazing electromagnetic properties and applications of artificial media, which are now known as metamaterials. There has been a significant amount of study done on the use of metamaterials in the development of miniature antennas that can overcome some of the disadvantages that are associated with the use of standard design methods.

High impedance surfaces (HIS) are man-made media that exhibit an exceptionally low reflection phase and, in certain circumstances, the suppression of all surface waves within a specified frequency range. This property distinguishes HIS from other types of artificial media. Typically, they are built from a dielectric substrate on which subwavelength metallic patches have been printed, and a metallic ground plane serves as the support for the structure. It is possible that the patches do not have any vias, or that the vias themselves are missing, which would result in a short between the metallic patches and the ground plane. The conductive channels that run between the metallic patches and the ground plane perform the function of an inductor, while the spaces that exist between the patches perform the function of an edge capacitor.

In addition to dosing a host dielectric with arrays of subwavelength metallic or dielectric impurities, one more option to construct metamaterials is to reactively load a transmission line with impurities. This method is known as reactive loading. In the one-dimensional configuration, a composite right-handed/left-handed transmission line medium can be created by periodically loading the host transmission line with series capacitors and shunt inductors. This medium allows for backward wave propagation in the low-frequency band and forward wave propagation in the high-frequency band.

6.3.1 Inventions of Antennas Based on Metamaterials

Antennas are said to be metamaterial-inspired when they interact with just one or a limited number of unit cells rather than an effective metamaterial medium. This type of interaction only occurs with metamaterial-inspired antennas. If it were arranged in an array according to the behavior of the unit cells, a metamaterial medium with either electrically negative (ENG), magnetically negative (MNG), or double-negative (DNG) properties would have the same material properties regardless of the type of negative property it possessed. The concept of antennas that are inspired by metamaterials is founded on the fact that the vast majority of unit cells that make up metamaterials are resonant structures. Unit cells are excited by the fields of an antenna and create a resonance at a frequency that is unique from the natural antenna resonance when they are positioned in close proximity to an antenna. Even though only one or a very small number of unit cells are used, the resonance of the metamaterial-inspired antenna frequently occurs at a frequency that is close to where the antenna would resonate in the presence of a metamaterial medium composed of an array of the identical unit cells. This is the case even though only one unit cell is used. Because the inclusion of the unit cells results in a resonance frequency that is likely to be much lower than that of the isolated antenna, the size of the isolated antenna is able to be lowered to the needed extent, meeting the requirements of the applications intended.

6.3.2 Miniaturization Using Metamaterials

The radiating patch, the ground plane, or the patch cavity (the gap between the patch and the ground plane) must be loaded with unit cell material in order to generate a patch antenna that is inspired by metamaterials.

For example, in the process of etching a unit cell pattern onto a patch, such as that of a split ring resonator (SRR), a complementary unit cell is produced. This complementary unit cell is activated when an electric field that is normal to the axis of the unit cell is applied to it. Due to the necessity of a regularly oriented excitation field, the optimal location for a complementary split ring resonator (CSRR) is either in the radiating patch, the ground plane, or the cavity of the patch itself. This method is analogous to the conventional way of adding slots to the patch or ground plane; however, in this instance, the slots are intended to function as resonant structures.

For example, Ortiz et al. demonstrates in [5] that a resonance at a frequency lower than the initial operating frequency of the patch can be created by etching out a single SRR from the radiating patch. This allows the resonance to occur at a frequency lower than the frequency at which the patch was originally designed to operate. Say, for example, in the beginning with a patch antenna designed to function at 4.8 GHz, the authors in [5] demonstrate that the size of the CSRR can be set to induce a resonance at 4.18 GHz by choosing to start with this antenna. The design makes use of a substrate that has a thickness of 0.49 mm and a dielectric constant of 2.43 at 4.18 GHz. This translates to $\lambda0/146$ at that frequency. At 4.18 GHz, the values for the radiation effectiveness, realized gain, and 10-dB fractional bandwidth are as follows: 17.92%, 0.11 dBi, and 0.72%, respectively. In comparison, the performance of the patch antenna at 4.8 GHz resonates at 64.83%, 5.85 dB, and 0.80%. In spite of the fact that a primary area reduction factor of 1/1.14 is achieved, the efficiency is significantly decreased. It is possible to further decrease the size of the patch antenna that was just described if the dimensions of the CSRR are altered or if different types of unit cells are used. Zhou et al. [6] discuss the design of a patch antenna that has a dimension of 0/9.5 and is resonant at 658 MHz. The antenna has a 10-dB bandwidth of 5 MHz (0.7%) and uses a complementary two-turn spiral resonator that is etched on the radiating patch. The resonance of the patch without the spiral resonator at a frequency of 2.75 GHz and a 10-dB bandwidth of 50 MHz (1.8%) implies a major area reduction factor of 1/4.18, which is a significant reduction in the total area. There is no information available regarding the effectiveness of the radiation. The normalized radiation pattern displays an almost 0-dB front to back ratio (FBR) at 658 MHz. This indicates that the antenna behaves more like a monopole than a directed patch antenna would. It has been demonstrated that increasing the number of CSRRs on the ground plane can increase both the multiband performance and the size. By adding two CSRRs to the ground plane of a patch that was initially resonant at 5.5 GHz, Xie et al. [7] demonstrate that it is possible to produce a second resonance at 3.87 GHz while still maintaining the initial resonance and the broadside radiation characteristics of the antenna. When compared to the antenna's realized gain and observed 10-dB fractional bandwidth at 5.5 GHz, which are, respectively, 2.7% and 6.2 dBi, the antenna's realized gain and observed 10-dB fractional bandwidth at 3.87 GHz are, respectively, 0.9% and 4.7 dBi. This results in a reduction of the bandwidth by 10 dB at the lower resonance, which is comparable to a reduction of the primary area by a factor of 1/1.42. In addition, the FBR has a decrease from around 15 dB at 5.5 GHz to 7 dB at 3.87 GHz.

The readings taken by the antenna at 3.87 GHz are as follows: $\lambda0/4.45 \times \lambda0/4.4 \times \lambda0/51.9$. It is feasible to greatly expand the preceding antenna's 10-dB bandwidth by adding additional CSRRs to the ground plane. Lee et al. illustrate in [8] how a patch operating at 4.3 GHz may have its resonance frequency changed to 2.96 GHz by covering the entire ground plane with an array of CSRRs. This allows the patch to operate at a lower frequency. The initial patch has a bandwidth of 2% at 4.3 GHz; however with CSRR loading, that bandwidth increases to 3.34% at 2.96 GHz. This results in a reduction of major area by a factor of 1/1.45, but it increases bandwidth by 1.67 times. The FBR is greatly reduced when using this strategy, which is a considerable drawback even for relatively moderate shrinking factors (to 0 dB). The prevalence of enormous slots in the ground plane is mostly to blame for this issue. These slots allow the antenna to behave less like a directed patch and more like an omnidirectional monopole. At 4.3 GHz, the measured realized gain is 6.05 dBi, which is significantly higher than the gain of 2.06 dBi at 2.96 GHz.

When metamaterial unit cells are placed into the patch cavity, it has also been proved that antenna size can be reduced. In [9], Ouedraogo et al. demonstrate that the surface area of the radiating patch and ground plane can be decreased by more than a factor of 1/16 while still maintaining a good impedance match with RL = 10 dB. This is accomplished by optimizing the geometry of a CSRR placed between the patch and the ground plane. This allows for a better impedance match to be achieved. In comparison to the unloaded patch, which has a radiation efficiency of 94 and a fractional bandwidth of 1.3%, the 1/16 area reduction results in a radiation efficiency of only 28 and a fractional bandwidth of only 0.4%.

The technology that is inspired by metamaterials has been applied to the process of miniaturizing several different types of antennas. A few examples of different types of antennas include the dipole, monopole, and loop types. The work in this area is the product of early conceptual studies on the usage of spherical ENG and MNG shells to realize tiny electric and magnetic antennas. These shells can be made of a variety of materials, including metal and nickel. As Ziolkwoski [10] and his team demonstrated, the inductive and capacitive properties of ENG and MNG spherical shells can be used to compensate for the capacitive and inductive properties of electrically tiny dipole and loop antennas. This can be accomplished by exploiting the properties of the shells. Theoretically, the dimensions of a dipole/loop antenna might be made infinitesimally smaller while yet maintaining a good impedance match, tremendous radiation efficiency, and wide bandwidth. This is something that could be done. The actual deployment of these ENG/MNG shells suffers from a number of problems, which are regrettably still present. The authors tried to develop electrically small dipole and loop antennas using a concept inspired by metamaterials.

References [11–19] likely describe different approaches to designing metamaterial-based antennas, such as using fractal geometries or planar structures to achieve miniaturization. These techniques have the potential to enable the development of compact and efficient antennas for a variety of applications, including mobile devices and IoT sensors.

6.4 WEARABLE ANTENNAS

In recent years, there has been a significant rise in the demand for wearable devices and the technology that is associated with them. This rise has been spurred by a number of significant breakthroughs, including the reduction in size of wireless devices, the development of high-speed wireless networks, the emergence of ultracompact, low-power SoCs, and the ongoing evolution of battery technology. These days, wearable electronics can be used for a broad variety of purposes, the majority of which rely on different types of antennas to sense, gather, and wirelessly transfer data with a host device or an IoT gateway.

6.4.1 What Exactly Are Wearable Antennas?

Wearable antennas are simply antennas that can be worn by the user. These antennas find widespread application in RF systems used in biomedicine as well as in wearable wireless communication devices. Wearable antennas are utilized within the framework of WBANs. Antennas are the primary component of a WBAN that is responsible for enabling wireless communication in all of its forms, including off-body, on-body, and in-body communication. Examples include GoPro action cameras, which have Wi-Fi and Bluetooth antennas and are frequently strapped to users in order to obtain their footage; smartwatches, which typically have built-in Bluetooth antennas; glasses, such as Google Glass, which have Wi-Fi and GPS antennas; and even the Nike+ Sensor, which connects to a smartphone via Bluetooth and is placed in a user's shoe.

A WBAN enables sensors, actuators, and IoT nodes to establish a wireless communication channel with one another on or beneath the skin, on clothing, or on the human body itself. Wearable antennas can be utilized by individuals of any age, athletes, and patients for the purpose of performing continuous monitoring of vital signs, oxygen level (oximetry), and stress level, among a variety of other parameters.

The development of high-efficiency, miniature antennas, which have many applications throughout the military, the medical field, and the consumer market, has made it substantially more viable to create both intrusive and noninvasive devices. A few examples of consumer-oriented wearable technology that use wearable antennas include smartwatches with built-in Bluetooth antennas, smart glasses with built-in Wi-Fi, GPS, and infrared (IR) antennas; body-worn action cameras with built-in Bluetooth and Wi-Fi, and tiny sensor devices in sports shoes with built-in Bluetooth and Wi-Fi that can connect to smartphones. A WBAN device allows for continuous monitoring of the health of a patient or senior citizen without interfering with their normal activities. Ophthalmic implants, cochlear implants, and heart pacemakers are only a few examples of the biological devices that make use of the implantable antenna sensors. Wearable antennas have a wide variety of applications in the military, including the tracking of a soldier's actual location, the transmission of real-time pictures and videos for decentralized communications that take place imme-diately, and many more. In addition to their usage in navigation, RFID applications, access and identity management, and other fields, these antennas have additional applications as well.

Because of their closeness to the human body, the wearable an-tennas in WBAN experience significant challenges, and vice versa. The impact that electromagnetic radiation has on the human body, the decreasing efficiency of the antenna as a result of the radiation pat-tern being fragmented, the changing impedance of the antenna, and the frequency detuning that occurs as a result of this fragmentation, call for special attention during antenna design for wearable devices.

When designing antennas for wearable technology, these spe-cific aspects require careful analysis and attention. When developing wearable antennas, designers need to pay particular attention to the structural deformation of the antenna, as well as the accuracy and precision of the production procedures, and the size of the antenna.

6.4.2   The Physiological Impact of Wearables on Humans

Even though nonionizing radiations like microwaves, visible light, and sound waves do not have enough energy to ionize the atoms or molecules in a body in the same way that ionizing radiations do, they are still able to raise cell temperature by moving or vibrating the atoms within the cell. This is because nonionizing radiations cause the atoms within the cell to move. Dielectric heating is a thermal consequence of microwave radiation that occurs when polar molecule rotations generated by the electromagnetic field heat a dielectric substance. This increase in temperature may cause catastrophic harm to human organs.

Antennas that do not have a ground plane have higher SAR values than those that do because the SAR of on-body antennas is dependent on the near field coupling to the body. As a consequence of this, altering the ground plane is an essential component of many different strategies for reducing the SAR value. One method involves the use of periodic conductive structures, such as electromagnetic bandgap (EBG) structures, for the purpose of filtering electromagnetic waves within certain frequency bands. Similarly, the application of HIS is helpful in obstructing the path of electromagnetic radiation that falls within a particular frequency band. The placement of wearable antennas in front of HIS enhances the front-to-back radiation ratio and reduces the SAR in the human body. In addition to this, HIS prevents surface waves from propagating and reflects electromagnetic waves without causing a phase shift. Another method that has proven to be effective is the utilization of an artificial magnetic conductor (AMC) ground plane, which performs the function of an isolator. SAR reduction strategies are often used by antenna designers, and common examples include incorporating metamaterials and ferrite sheets.

6.4.3   The Effect of the Human Body on Antennas That Are Worn

Wearable antennas are susceptible to the effects of the human body when it is in close proximity to them. The human body has features that cause it to have a high loss factor and a high dielectric constant.

These properties can induce changes in input impedance, as well as shifts in frequency and a reduction in antenna efficiency. The communication route that is supposed to run between the antenna and the external host device has been messed up.

Depending on the particular use case, one can choose from a number of different strategies to mitigate the effect that the human body has on antennas. One of the most significant considerations is the location of the antenna in addition to its orientation. The optimal position, orientation, and distance of an antenna in relation to the human body significantly mitigates the impact of the body on the performance of antennas. In high-performance systems, you can also find applications for automatically tunable circuits and programmable antennas. Wearable antennas are designed using HIS and EBG ground planes in order to reduce the impact of the human body on the performance of the antennas.

6.4.4   Conclusion

Wearable antennas are quickly becoming one of the most significant emerging technologies. These antennas have a wide range of applications, including those in the sectors of entertainment, navigation, medical care, and the military. With the use of WBAN technology, in particular wearable antennas, it is possible to remotely sense and monitor a variety of physiological aspects of the human body. Wearable antennas are particularly useful for this purpose. When considering the advantages of WBAN and wearable antennas, it is essential to keep in mind the potential effects that these technologies may have on human physiology. Antenna designers are the ones who should decide whether RF technology is suitable for wearable designs so that there is the least amount of negative impact on the device efficiency and gain caused by electromagnetic immersion in human tissue.

When there are so many different kinds of antenna technologies available, it might be challenging to pick the one that is best suited for a particular wearable application. In a manner comparable to this, antenna design is a difficult process that calls for the use of complex simulation tools in conjunction with skilled RF antenna designers. The most effective technique for limiting the risk is to create as much space as possible between the antenna and the body. This is something that should be rather clear, but it is also really crucial. It is important to note that GPS antennas that are attached to the body have certain restrictions to adhere to. Antennas for GPS, GNSS, and GLONASS are typically constructed to have a wide field of view of the sky above them so that they can detect a large number of satellites. When used as a wearable antenna, the line of sight of antenna is partially obscured because the body blocks approximately half of it. Because of this, achieving satisfactory performance with GPS on wearable technologies is incredibly challenging.

6.5 CONFORMAL ANTENNAS

As the technology behind communication and navigation continues to advance at a rapid pace, the implementation of conformal antennas is becoming increasingly important for enhancing the transfer and reception properties of antennas. The term *conformal antenna* refers to an array that may be molded to conform to a specific shape. The antenna shape is selected for reasons other than electromagnetic compatibility. As the communication and route frameworks of flying machines change, new cutting-edge advances are generated, and new methods are required to manage the components of the framework. The conformal antenna construction may take the form of a barrel shape, a circular shape, a tapering shape, or another geometry entirely, depending on the curvature of the surface it is attached to. Nevertheless, this geometry suffers from a number of limitations, the most notable of which are its complicated design and its challenging compatibility with the existing radiation pattern. In conformal antennas, the area of transmission and reception is significantly larger than it is on planar surfaces. As a result, conformal antennas provide a broad beam radiation pattern that covers a larger region than other surfaces. Conformal combination is required since there is an increase in the number of sophisticated and dedicated antennas in vehicle frameworks, versatile communication, and airborne and spaceborne applications respectively. The term conformal antennas is still used to refer to antennas that are connected to the curved surface of a mechanical bearer. Even though the principal forms that have been examined up to this point are funnel-shaped, circular, or round and hollow, conformal antennas can be mounted on any geometric structure. In order to travel at great speeds, rockets require antennas that are both thin and conformal. A reception apparatus that may be described as paper thin would be appropriate for the aerodynamic and mechanical specialist.

The advantages listed as follows are ones that conformal antennas have over planar tiny antennas:

- Provides coverage from a variety of perspectives;
- The installation of radomes is not required, thus there will be no need to account for the slight increase in losses caused by radomes. The enhancement of the aerodynamic profile, the omnidirectional radiation pattern given by cylindrical conformal antennas, and the capability of conformal antennas to span 360 degrees are all benefits of using these antennas.

The following are some of the problems that conformal antennas have, which are caused by bending and bedding:

- The substrate material is stretched, which results in a little inaccuracy.
- As a result of bending, the substrate material may crack, which has the potential to impact the performance of the entire structure.

Planar antennas have been the subject of investigation by a great number of researchers; nevertheless, these devices are constrained in a number of ways, including their inability to conform to curved surfaces and the requirement that they produce additional aerodynamic drag. The conformal antenna is able to provide a consistent, wide-ranging, and exact scope, and it may be put on a surface that is either rounded or hollow. The primary concentration of the examination throughout its entirety has been on the conformal antenna in the shape of a tube. Increasing the substrate height, cutting slots, relocating the feeding point, adopting layered substrates, split-ring antennas, array antennas, and other designs have all been proposed as potential methods for raising the gain and the bandwidth of an antenna. By utilizing metamaterial, it is possible to boost the gain and bandwidth of the antenna while still keeping its good radiation properties. Structures such as SRRs and CSRRs are utilized on a regular basis for the purpose of determining whether or not a substrate material possesses metamaterial properties. The many designs of SRR antennas that are now available on the market are subject to a variety of constraints, including those pertaining to size, gain, and bandwidth.

According to the International Electrotechnical Commission (IEC), a cavity antenna (CA) is a radiating system whose shape is not dictated by its electromagnetic properties but rather by the surface of the system where it must be absorbed. This information comes from the IEC. When developing a system, it is vital to make use of new methodologies and strategies due to the rapid advancement of technology. The system that acts as an interface between the transmitter and receiver while they are in free space is what is meant to be understood by the term *typical radiating system*. It is possible to alter the properties in order to improve the overall functionality of the system as a whole. This approach frequently lowers a number of aspects that have a large negative impact on the picture metrics, which ultimately leads to improved accuracy, outstanding aerodynamics, and a reduction in volume. As a consequence of this, the antenna engineers face a challenging task in terms of the construction of conformal structures rather than planar ones. This antenna can be mounted on a wide variety of aerodynamic systems because of its adaptability. In addition to that, they must have conformal properties and possess multiband, broadband, and miniaturization capabilities. The CA is a phased array antenna of a particular type. The CAs are made up of a surface-covering array comprising a large number of small, flat antennas that are identical to one another, such as dipoles and patches. Each and every antenna possesses a phase shifter that is controlled by a computer and works by merely shifting the phase of the current that is being applied to each element. It is feasible to combine all of the radiation into a powerful beam that is aimed in a particular direction by making use of interference as the connecting mechanism. On the other hand, every part of the receiving system combines all of these waves into phase-independent signals traveling in that direction. Because of this, the antenna will be able to respond to the signal originating from a single transmitter while simultaneously reflecting any conflicting signals coming from other directions. In contrast to conformal antennas, which have their elements dispersed across a curved surface, ordinary phased arrays have their elements distributed across a flat surface. Matching phase shifters are employed in order to correct the phase discrepancy that occurs as a direct consequence of varying path lengths. Due to the miniature nature of the CA elements in question, these applications can only be carried out at high frequencies and within the microwave spectrum. These CAs are characterized by lesser visibility and are capable of being incorporated on the structure as a result of their inherent miniaturization. This quality is absolutely essential in a military situation. The CA is compatible with any geometry you choose to use.

When compared to planar antennas, the microstrip patch antenna for cylindrical surfaces possesses superior radiation efficiency, more directivity, and a lower antenna diameter. Additionally, the microstrip patch antenna for cylindrical surfaces is more compact. In comparison to other types of constructions, the cylindrical shape offers low levels of cross polarization and has the potential to have a lower value for the value of the standing wave ratio. In this scenario, the use of a greater number of antennas will make it possible to generate an omnidirectional pattern that has high gain values and a bandwidth range that is appropriate for its needs. Circular polarization can be produced with the application of an original feeding strategy. A dipole-like radiation pattern is produced in the vertical plane by the structure that is circularly cylinder-shaped. This structure offers near to 360°-coverage in the horizontal plane. The spherical surface antenna array provides for simple and effective regulation of the phase excitation.

According to the findings of the literature available, spherical microstrip patch antennas give a better main lobe pattern and a side lobe pattern that is less apparent than planar surfaces. The hemispherical antenna array not only boosted bandwidth but also had a more favorable amplitude profile, and it was responsible for providing dual linear polarization. The beam forming network is able to generate a broadband radiation pattern by making some adjustments to the conventional least squares synthesis. Microstrip patch antennas that have a shape that is both spherical and cylindrical can be used in the construction of conformal antennas. Because of this, the curved surface provides improved performance in terms of antenna parameters such as gain, bandwidth, beamwidth, and loss reductions. In addition, when these geometries are utilized as conformal antennas, the amount of cross polarization that occurs is reduced and increases the bandwidth range. According to the findings of the literature available, on the dielectric substrate, it is recommended to make use of a substrate that has a high relative permittivity in order to generate a broad beam radiation pattern. Additionally, the use of metamaterials helps to reduce the amount of mutual coupling that exists between arrays as well as the overall size of the radiator. On the other hand, it continues to thicken and becomes increasingly difficult to bend. As a result, the design of a conformal antenna calls for a significant amount of concentration.

6.6 CONCLUSION

Miniaturization technologies have become an essential aspect of modern communication systems. The advent of portable electronic devices and the need for more efficient communication systems in various sectors have necessitated the development of smaller antennas with high performance. Practicalities in miniaturization involve various considerations such as the materials used, fabrication techniques, and performance trade-offs.

However, miniaturization also poses performance limitations for tiny antennas such as reduced bandwidth and gain. Metamaterial-based miniaturization has emerged as a promising solution to these challenges, allowing for the creation of compact antennas with high performance. Inventions of antennas based on metamaterials have led to various applications such as wearable antennas, which have become increasingly popular due to their potential for health monitoring and other physiological impacts on humans.

The use of metamaterials in miniaturization has also led to the development of conformal antennas, which can be integrated into various objects and surfaces while maintaining their performance characteristics. These antennas offer unique advantages in terms of flexibility and versatility, making them suitable for various applications, including military, aerospace, and healthcare.

Overall, miniaturization technologies and the use of metamaterials have opened up new opportunities for the development of compact and high-performance antennas that can be integrated into a wide range of applications. These advancements will continue to shape the future of communication systems and the way we interact with technology.

References

[1]   Wheeler, H. A., "Fundamental Limitations of Small Antennas," *Proc. IRE,* Vol. 35, 1947, pp. 1479–1484.

[2]   Chu, L. J., "Physical Limitations of Omnidirectional Antennas," Technical report, No. 64, MIT Research Laboratory of Electronics, May 1, 1948.

[3]   Chu, L. J., "Physical Limitations of Omnidirectional Antennas," J. Appl. Phys., Vol. 19, 1948, pp. 1163–1175.

[4]   Balanis, C. A., *Antenna Theory: Analysis and Design*, 4th edition, Hoboken, NJ: John Wiley & Sons, 2016.

[5]   Ortiz, N., F. Falcone, and M. Sorolla, "Dual Band Patch Antenna Based on Complementary Rectangular Split-Ring Resonators," in *Proceedings Asia-Pacific Microwave Conference (APMC2009)*, Singapore, 2009, pp. 2762–2765.

[6]   Zhou, L., S. Liu, Y. Wei, Y. Chen, and N. Gao, "Dual-Band Circularly-Polarized Antenna Based on Complementary Two Turns Spiral Resonator," *Elect. Lett.*, Vol. 46, 2010, pp. 970–971.

[7]   Xie, Y-H., L. Li, C. Zhu, and C-H. Liang, "A Novel Dual-Band Patch Antenna with Complementary Split Ring Resonators Embedded in the Ground Plane," *Prog. Electromagn. Res. Lett.*, 2011, Vol. 25, pp. 117–126.

[8]   Lee Y., and H. Yang, "Characterization of Microstrip Patch Antennas On Metamaterial Substrates Loaded with Complementary Split-Ring Resonators," Microw. Opt. Tech. Lett., Vol. 50, 2008, pp. 2131–2135.

[9]   Ouedraogo, R. O., E. J. Rothwell, A. R. Diaz, K. Fuchi, and A. Temme, "Miniaturization of Patch Antennas Using a Metamaterial-Inspired Technique," *IEEE Trans. Ant. Propag.*, Vol. 60, 2012, pp. 2175–2182.

[10] Erentok, A., and R. W. Ziolkowski, "An Efficient Metamaterial-Inspired Electrically-Small Antenna," *Microw. Opt. Tech. Lett.*, Vol. 49, 2007, pp. 1287–1290.

[11] Lim, J. S., et al., "Design of a Subwavelength Patch Antenna Using Metamaterials," in *38th European Microwave Conference (EuMC)*, Amsterdam, 2008, pp. 1246–1249.

[12] Ouedraogo, R. O., and E. J. Rothwell, "Metamaterial-Inspired Patch Antenna Miniaturization Technique," in *IEEE AP-S International Symposium and National Radio Science Meeting*, Toronto, ON, 2010, pp. 1–4.

[13] Jahani, S., J. Rashed-Mohassel, and M. Shahabadi, "Miniaturization of Circular Patch Antennas Using MNG Metamaterials," *IEEE Ant. Wireless Propag. Lett.*, Vol. 9, 2010, pp. 1194–1196.

[14] Pruitt, J., and D. Strickland, "Experimental Exploration of Metamaterial substrate Design for an Electrically Small Patch-like Antenna," in *IEEE AP-S International Symposium and National Radio Science Meeting*, Toronto, ON, 2010, pp. 1–4.

[15] Ouedraogo, R. O., E. J. Rothwell, A. R. Diaz, K. Fuchi, and A. Temme, "Miniaturization of Patch Antennas Using a Metamaterial-Inspired Technique," *IEEE Trans. Ant. Propag.*, Vol. 60, pp. 2175–2182.

[16] Tang, M.-C., and R. W. Ziolkowski, "A Study of Low-Profile, Broadside Radiation, Efficient, Electrically Small Antennas Based on Complementary

Split Ring Resonators," *IEEE Trans. Ant. Propag.*, Vol. 61, 2013, pp. 4419–4430.

[17] Paul, P. M., et al., "Miniaturization of Square Patch Antenna Using Complementary Split Ring Resonators," in *Advances in Computing and Communications (ICACC), Third International Conference*, Cochin, 2013, pp. 122–125.

[18] Dai, X.-W., Z.-Y. Wang, L. Li, and C.-H. Liang, "Multi-Band Rectangular Microstrip Antenna Using a Metamaterial-Inspired Technique," *Prog. Electromagn. Res. Lett.*, Vol. 41, 2013, pp. 87–95.

[19] Kamil, B. A., D. C. Mehmet, B. Filiberto, T. Alessandro, V. Lucio, and K. Ekmel, "Experimental Verification of Metamaterial Loaded Small Patch Antennas," *COMPEL: Int. J. Comput. Math. Elec. Elect. Eng.*, Vol. 32, 2013, pp. 1834–1844.

7

IoT IN SMART CITIES

7.1 INTRODUCTION

A smart city is a municipality that makes use of information and communication technologies (ICT) to boost administrative effectiveness, disseminate information to the general public, and enhance the standard of public services as well as the welfare of its residents. While the precise definition varies, a smart city's overarching goal is to use smart technology and data analysis to optimize city operations, spur economic growth, and enhance residents' quality of life. The worth of a smart city is determined by how it chooses to use its technology, not just how much of it the city may have.

A city's smartness is assessed using a number of key factors. These qualities consist of:

- A network based on technology;
- Environmental programs;
- An efficient network of public transportation;
- An assured understanding of urban planning.

The success of a smart city rests on its capacity to forge a solid alliance between the public and private sectors, notably with regard to bureaucracy and rules. The majority of the labor required to establish and sustain a digital, data-driven environment is done outside of the government, thus this partnership is essential. Sensors from one business, cameras from another, and a server from a third company might all be part of the surveillance apparatus for crowded streets.

Additionally, it is possible to employ independent contractors to study the data and then submit their findings to the local government. This information could then influence the hiring of an application development team to incorporate a remedy for the issues identified in the examined data. If the solution needs ongoing monitoring and updates, this business might join the system. As a result, developing a successful network of partnerships rather than finishing a single project becomes more important for the success of a smart city.

7.2   WHY ARE SMART CITIES NECESSARY?

The main objective of a smart city is to develop an urban setting that offers its citizens a high quality of life while also fostering general economic growth. As a result, a key benefit of smart cities is their capacity to enable increased service delivery to inhabitants with less infrastructure and expense.

In order to handle the expanding population, metropolitan communities must utilize their infrastructure and resources more effectively as the population of their cities keeps growing. Applications for smart cities can make these changes possible, boost city operations, and enhance resident quality of life. Applications for smart cities let cities extract additional value from their existing infrastructure. The upgrades enable additional revenue sources and improved operational effectiveness, which saves money for both the government and the populace.

Traffic management, sustainable infrastructure, safety and security, and digital services are the four key application areas for smart cities as depicted in Figure 7.1.

Smart parking and smart intersections are part of traffic management, where communications are provided to traffic signals to permit the movement of traffic more quickly through the city. These decisions are made in real time utilizing traffic monitoring and the

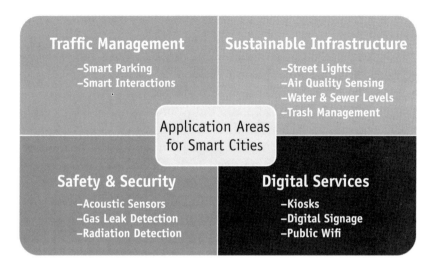

Figure 7.1 Application areas of smart cities.

amount of traffic data. The theory is that one can reduce congestion if one can signal it just right and circulate the traffic across city blocks.

Streetlights fall under the second category of sustainable infrastructure, which allows for less energy use and a smaller carbon imprint. It enhances quality of life from a safety standpoint. Other examples of sustainable infrastructure include air quality sensing, which identifies areas with greater pollution levels. Then there are water levels; identifying probable flood zones depends on where one might have a high water level. Similarly, sensors for sewer levels can give proactive alerts and warnings to dispatch teams to solve infrastructure issues before they arise. No one likes to think about when sewers overflow and pose potentially serious safety risks.

Another illustration is clever trash management. One can send sanitation staff to empty waste bins only when they are full by installing sensors on the bins. This enables businesses to operate more cheaply. Smart trash management also relates to traffic management since it prevents vehicles from having to go empty a container that is just partially filled.

Security and safety comprise the third category. For example, acoustic sensors for gunshot detection could fall within this category. Sensors will detect the sound and determine if it is a gunshot or simply a car backfiring. The shooter's location can then be ascertained and sent to public safety officials utilizing triangulation techniques.

Gas leak detection is another illustration. With Con Edison in New York City, methane gas detectors have been installed that can detect possible leaks before they become a problem. Another example is the detection of radiation in and out of houses.

Digital services, which include items like kiosks and digital signage, make up the final category. Digital adverts bring in money for the city, but they can also be used to broadcast public service announcements (PSAs), such as Amber or Silver Alerts for children or people who have gone missing. Wayfinding is another purpose for kiosks. The kiosk provides information on neighboring companies and sites of interest for city visitors. Public Wi-Fi is an example of a digital service.

The IoT gadgets, software programs, user interfaces (UI), and communication networks are all utilized by smart cities. They do, however, heavily rely on the IoT. The IoT is a network of interconnected items that can exchange data and communicate, including cars, sensors, and home appliances. The cloud or servers are used to store the data that IoT sensors and devices collect and deliver. The interconnection of these devices and the use of data analytics (DA) enable the convergence of the physical and digital city components, enhancing the effectiveness of the public and commercial sectors, generating economic gains, and enhancing the lives of citizens.

Edge computing is a processing capability that some IoT devices have. Only the most crucial and pertinent information is transmitted via the communication network thanks to edge computing. The protection, supervision, and management of network traffic within a computing system also need the use of a firewall security system. By blocking any illegal access to the IoT network or city data, firewalls maintain the security of the data that is continually being transmitted inside a smart city network.

Other innovations for smart cities are listed next.

7.2.1 Interfaces for Application Programming

- Computerized intelligence (AI);
- The use of clouds;
- Dashboards;

- Computer learning (ML);
- Intermachine (M2M);
- Network mesh.

7.2.2 Qualities of Smart Cities

The adoption of smart cities is being fueled by new trends like automation, machine learning (ML), and the IoT. Theoretically, a smart city effort might include any aspect of city administration. The smart parking meter, which uses an app to assist cars in finding available parking spaces without needlessly circling busy city blocks, is a classic example. Digital payment is also supported by the smart meter, so there is no worry of running out of coins for the meter.

Smart traffic management is used in the transportation industry to track and analyze traffic flows in order to improve streetlights and keep roads from getting too crowded during peak hours. Another aspect of smart cities is smart public transportation. Real-time service coordination and customer happiness are made possible by smart transit firms, which increase productivity and satisfaction. In a smart city, bike- and ride-sharing are both prevalent services.

Smart cities place a lot of emphasis on efficiency and energy conservation. Smart streetlights use smart sensors to dim when there aren't any vehicles or people on the roads. Smart grid technology can be used to monitor energy outages, supply electricity on demand, and optimize operations, maintenance, and planning. Initiatives toward a smarter city also seek to track and mitigate environmental issues like air pollution and climate change. Smart technology can also be used to improve waste management and sanitation, whether it be through the use of internet-connected trash cans and IoT-enabled fleet management systems for waste collection and removal, or through the use of sensors to measure water parameters and ensure the quality of drinking water at the system front end with proper wastewater removal and drainage at the system back end.

By monitoring high-crime areas and using sensors to improve emergency preparedness, smart city technology is being used more and more to increase public safety. Smart sensors, for instance, can play a key role in early warning systems for hurricanes, landslides, floods, and droughts. A smart city project frequently includes smart buildings as well. Sensors can be integrated into new construction and retrofitted into existing infrastructure to enable real-time space management, maintain public safety, and monitor the structural health of buildings. When repairs are required, sensors may spot wear and tear and alert the appropriate authorities. Using a smart city application, citizens can alert authorities when repairs are required for buildings and other public infrastructure, such as potholes. In order to lower expenses and increase worker productivity, sensors can also be used to find leaks in water mains and other pipe systems.

7.3   HOW SMART CITIES OPERATE

Smart cities use their network of interconnected IoT devices and other technology to improve livability and spur economic development. Four steps are followed by successful smart cities:

1. Data collection: Real-time data is gathered by smart sensors placed all around the city.
2. Analysis is the process of evaluating the data gathered by the smart sensors in order to derive useful conclusions.
3. Communication: Through effective communication networks, the insights discovered during the analytical phase are shared with decision makers.
4. Cities take action by utilizing the information gleaned from the data to develop solutions, enhance operations and asset management, and raise resident quality of life.

7.4   SMART CITY SUSTAINABILITY PROMOTION

Another important aspect of smart cities is sustainability. In the upcoming years, urbanization is predicted to rise even further. According to the United Nations, around 55% of the world's population currently lives in a city or metropolitan region, and over the next few decades, that percentage is expected to increase to 68%. In the coming years, smart technology will assist metropolitan regions in maintaining growth while enhancing efficiency for both citizen welfare and governmental effectiveness.

While cities already have some environmental benefits, such as smaller ecological footprints that have less of an impact, they also have negative environmental effects due to emissions, such as the excessive use of fossil fuels. The network of technology for smart cities may be able to mitigate these negative impacts [1].

Switching to an electric public transportation system would reduce gasoline emissions while also having the benefit of collaborating closely with the city's electric power infrastructure to reduce the impact of battery charging during periods of high electric demand. Additionally, while not in use, electric vehicles might be utilized to control the frequency of the city's electric grid with the right coordination.

As cities become more intelligent, it is also anticipated that fewer cars would be used in urban areas. Self-driving automobiles or autonomous vehicles [2] have the potential to alter human perception of how necessary it is to own a car. It is predicted that the use of autonomous vehicles would lead to a decrease in the number of vehicles owned by ordinary people, which will lead to a reduction in the number of automobiles on the road and the emission of harmful gases.

7.5   CHALLENGES AND CONCERNS WITH SMART CITIES

Initiatives for smart cities must involve the people it seeks to assist: locals, business owners, and tourists. City officials must encourage the usage of open, democratized data among their constituents as well as raise knowledge of the advantages of the smart city technology being adopted. People are more willing to participate if they are aware of what they are doing and the potential advantages.

The key to producing a smart citizen who will be engaged and empowered to positively contribute to the city and community is to foster collaboration between the public and private sector and local citizens. Plans for making the data public and accessible to citizens, frequently through an open data portal or mobile app, should be included in smart city projects. Residents can then interact with the data and learn how it is used. Residents may also be able to perform personal tasks using a smart city app, such as viewing their home energy usage, paying bills, and locating effective public transportation.

Opponents of smart cities believe that municipal officials won't prioritize data privacy and security, for fear that doing so exposes citizen data to the risk of hacking or misuse. Furthermore, the existence of sensors and cameras could be interpreted as a form of governmental surveillance or a breach of privacy. Data from smart cities should not contain personally identifying information and should be anonymized in order to address this.

The connectivity issue, however, may be the biggest difficulty that smart cities must overcome. Without a strong connection, the hundreds of thousands or millions of IoT devices dispersed around the city would stop working, and the smart city would cease to exist.

More so when a system ages and expands, public transportation, traffic control, public safety, water and waste management, electricity supply, and natural gas supply can all be unreliable. However, as the city grows and the demands placed on its infrastructure rise, the significance of these processes will only grow. To make sure that they are operating properly, these systems need to be tested and maintained continuously.

Finding strategies to draw and keep citizens in communities with a strong cultural fabric presents another problem for smart cities. Residents are frequently drawn to a place by its unique culture, which cannot be encoded into a sensor or controlled by one. Additionally, new smart cities that are being developed from the ground up [3], such as Neom in Saudi Arabia and Buckeye in the desert-region of Arizona, face the challenge of having to attract citizens because they lack an existing population. These potential smart cities also lack any track record of accomplishment to inspire confidence. There have been questions raised over whether a sustainable water source is even accessible as Neom and Buckeye have been constructed.

7.6   WHO AND HOW TO USE DATA FROM SMART CITIES

There are several audiences. One is the city or utility itself, whose crews are in charge of upkeep on the lighting system. The information on alerts and failures is mostly intended for them. Similarly, traffic departments use this information to decide whether to close roads or expand them.

Analysts and consultants who provide tools that can consume this data are a new audience that has begun to emerge. Instead of flooding cities with data sets, their algorithms help users make judgements.

The idea of open data [4] is another. For instance, a client in the United Kingdom had a strong connection to the community and university there. To promote innovation in the city, any data coming in from the network was placed into an open data platform. This allowed developers to use the data to create the next traffic management software or app by approaching it differently than others have in the past. Each city agency typically only considers its own demands and possible solutions, which is a concern. Multiple networks and applications sometimes provide the same function in each city as a result. From an IT standpoint, this results in higher overhead expenses for management and maintenance. By using a single network architecture, cities and their departments may become smarter. There is a missed potential for additional savings when one has all these different data sets and is not combining them.

One could combine information on noise, lighting, and traffic and ask, "OK, what can we infer from this data set? Do we need to adjust the timing or sequencing of the traffic light as a result?" A shared data platform, where one has the chance to cross-reference the data sets, is essential to this.

7.7   THE ROLE OF SMARTPHONES IN IoT-ENABLED SMART CITIES

The smartphone is IoT for the common individual since it will be available everywhere and everyone carries one around all day. Smartphones are the gateway to the IoT. We use our smartphones to control smart devices in our homes, cars, and workplaces. This is possible because smartphones are always with us and have the computing power and connectivity to communicate with other devices. If we are worried that something is wrong with our IoT device (for example, if the check engine light on our own car comes on), one can use a professional vehicle scan diagnostic tool from one's smartphone to read diagnostic trouble codes, which is less expensive than taking it to a qualified automechanic to diagnose the issue. Additional options include linking a smartphone to a washing machine to view a graph of water and electricity usage over the previous several weeks or months. Generally speaking, every IoT item, including refrigerators, vehicles, and washing machines, may connect to and exchange data using smartphones, with specially configured servers on the internet.

7.8   PRACTICAL ASPECT OF SMART CITY AND ITS COMPONENTS IN THE IoT ERA

Tokyo, the metropolis with the highest population density in the world, continues to expand and has the most inhabitants of any city on Earth. With a population of more than 38 million, the capital city of Japan is the biggest city in the world (38,050,000 people). In addition, there are over 26 million people living in Delhi, India, and over 31 million (32,275,000) people reside in Jakarta, Indonesia. By 2030, major cities will be home to 60% of the world's population, according to projections [5].

The results include a lack of freshwater, a waste mountain, a traffic jam, and air pollution. How can we overcome these obstacles? A smart city, a networked and intelligent city, is one of the keys. It speaks for a higher standard of living and less resource consumption.

Here are five elements of the smart city and how they will affect the IoT:

1. *Ingenious infrastructure*: Cities must establish the circumstances for ongoing development, with urban infrastructure and buildings needing to be developed more sustainably and effectively as digital technologies become more vital. CO_2 emissions should be reduced to a minimum by, for instance, purchasing electric automobiles and self-propelled vehicles. To achieve an energy-efficient and ecologically friendly infrastructure, smart cities deploy sophisticated technologies [6]. To limit the demand for electrical power, smart lights should only turn on when someone really walks past them by controlling brightness levels and monitoring daily use.

2. *City Air Management (CyAM) instrument*: Siemens has created a comprehensive cloud-based software suite called the CyAM tool, which anticipates emissions and captures pollution data in real time. Forecasts for the upcoming three to five days' worth of emissions can be made with an accuracy of up to 90%. The CyAM tool is unique because it predicts air pollution and measures its effectiveness and the technologies that are utilized. An algorithm that utilizes an artificial neural network underlies the prediction. CyAM is a cloud-based software package that includes a dashboard that shows real-time data on the air quality identified by sensors throughout

a city and forecasts values for the following three to five days. To simulate the future three to five days, cities can select from 17 metrics (effects of the air quality for the upcoming three to five days).

Siemens' cloud-based, open operating system for the IoT, MindSphere, serves as the foundation for CyAM (IoT).

3. *Traffic control*: Optimizing traffic in big smart cities is a challenge. Los Angeles, one of the world's biggest cities, has installed an intelligent transportation system to manage traffic flow. Real-time updates of traffic flow are sent by sensors incorporated into the pavement to a centralized traffic management platform that analyzes the information and instantly adapts the traffic signals to the situation. It predicts possible traffic patterns using previous data, with no human intervention.

4. *Adaptive parking*: Smart parking technologies can detect when a car has left the parking space. The sensors in the ground notify the vehicle via smartphone of available parking spaces that are free. Others employ vehicle input to pinpoint the locations of the openings and direct awaiting vehicles in the direction of the path of least resistance. Smart parking is a reality today and, because it doesn't need a complex infrastructure or a lot of money, it's perfect for a midsized smart city [7].

5. *Waste management done right*: Waste management systems aid in improving waste collection efficiency, lowering operational costs, and more effectively addressing the environmental problems [8] brought on by ineffective waste collection. A level sensor is attached to a waste container, and when a specified level is reached, a truck driver's management platform notifies him via his smartphone. To prevent drains from becoming partially full, the message seems to empty a full container.

7.9   FUTURE IoT USE CASES

Here is a look at some cutting-edge IoT gadgets.

1. *Mercedes-Benz Vision Van*: Mercedes-Benz's Vision Van is a van design for metropolitan areas that has various cutting-edge technologies, including an autonomous drone delivery system. Within a 10-km radius, the drone can deliver on its own. The parcel delivery service would also benefit from time savings during loading and delivery. In contrast, one-shot loading just needs around five minutes, while manual loading can take up to an hour and a half. The automation technology in the hold and the drones delivering in parallel to the deliverer reduce the last mile delivery time by up to 50%.

2. *Shrewd eye*: This smart eye technology is quite similar to Google Glass, its most ambitious effort. In order to present alternatives and accessible features right in front of your eyes without being distracting, the smart eye is outfitted with sensors, Wi-Fi, and Bluetooth. It is now feasible to read texts, browse the internet, and more thanks to technology.

7.10   IoT SENSORS

IoT today frequently makes use of sensors [9] which play an increasingly important role in modern city planning, as they provide a wealth of data that can be used to improve the efficiency and sustainability of urban environments. They are present everywhere you walk, including in buildings, schools, and IoT-enabled smart cities. You might not even notice them. In order to communicate data with other linked devices and management systems, sensors can be employed in a range of sources, including heat, pressure, water, and motion. Following are the current trends in sensor design:

1. *Miniaturization*: Sensors are becoming smaller and more compact, with the ability to fit into smaller devices and equipment. This trend is driven by the need for sensors to be integrated into wearable devices and other portable technologies, such as smartphones and smartwatches. For example, the Apple Watch features a variety of sensors, including an

accelerometer, gyroscope, heart rate sensor, and electrocardiogram (ECG).

2. *Wireless connectivity*: Sensors are increasingly incorporating wireless connectivity options, allowing for real-time data transmission and remote monitoring. This trend is particularly evident in the field of healthcare [10], where sensors can monitor vital signs and transmit data to medical professionals for remote diagnosis and treatment. For example, the KardiaMobile device is a small, wireless ECG monitor that connects to a smartphone app, allowing patients to monitor their heart health from home.

3. *Integration of AI and ML*: Sensors are being designed to integrate with AI and ML algorithms, allowing for more advanced data analysis and predictive modeling. This trend is particularly evident in the field of industrial automation, where sensors are used to collect data on manufacturing processes and machine performance. For example, sensors on an assembly line can collect data on production rates and identify areas for improvement, with ML algorithms providing recommendations for optimization.

4. *Improved energy efficiency*: Sensors are becoming more energy-efficient, allowing for longer battery life and reduced environmental impact. This trend is driven by the need to minimize power consumption in portable devices and IoT applications, where sensors must operate on limited battery power. For example, the Nordic Thingy:91 is a multisensor cellular IoT prototyping kit that features ultralow power consumption, allowing for up to five years of battery life.

The advancements are enabling sensors to be integrated into a wide range of applications, from wearables and smartphones to industrial automation and IoT devices. Urban farming and manufacturing are also made more efficient by smart city technologies, which also creates jobs and improves space management and provides consumers with fresher items that are more accessible.

7.10.1 Sensors for Temperature

Uses for temperature sensors include:

Agriculture: Temperature sensors can be used to monitor the soil's and the atmosphere's temperatures. They make agriculture management more effective and are practical for farmers. Farmers may control when and how much water is applied to their crops, as well as predict future conditions using data.

Cold storage: To support and maintain suitable environmental conditions, temperature sensors are needed. In blood bank facilities, medical labs, and vaccine storage, this device is frequently used since we can set the highest or lowest temperature, and it may also alarm us.

Food industries: IoT systems can use sensors to keep food secure while being transported in refrigerated vehicles. The temperature can be displayed in real time, and data can be stored for later retrieval. These will aid in raising the level of service quality and dependability.

Humidity gauges: The amount of water vapor in an atmosphere of air or other gases is measured using humidity sensors. The percentage of relative humidity (RH) serves as the unit. The relative humidity can be measured by the humidity sensor between 10% and 90% RH. It is typically used in conjunction with a temperature sensor, but because of its tiny size and low cost, it is frequently utilized in homes and businesses.

Humidity sensors are essential tools for managing the production process in a variety of industries.

Paper industry: Humidity sensors are used for humidity management, which helps improve the effectiveness of the drying process in the pulp and paper business, sheet moisture control, and can aid in the diagnostics process. Temperature and moisture must be present for food production. The flavor may be impacted. Devices like dishTemp thermometers, Food Check thermometers, and frying thermometers are needed by the food sector to regulate standards. These devices all employ humidity sensors to look for moisture.

Clean room/vaccine storage: A clean room or vaccine storage space calls for particular security because it must adhere strictly to a number of regulations, including those relating to pressure, temperature, humidity, and wind. The development of microbes, bacteria, dirt, or pollutants will be impacted by these conditions. Therefore, the humidity sensor [11] is essential in many places, including hospitals, laboratories, and research facilities, among others.

Agriculture: Humidity sensors are used to detect water or to measure soil moisture. It can communicate with the microcontroller using either the digital signal supplied by the module or an analog readout of the moisture level. With the use of these sensors, you may create an automatic irrigation system that can plant your crop for you so that you don't have to.

7.10.2 Gas Detectors

The development of toxic, flammable, or dangerous gases as well as the depletion of oxygen are among the air quality changes that are detected by a gas sensor. Due to its application in homes, we are perhaps most familiar with this sensor. It is frequently used in manufacturing, chemical research, oil, and mining. Gases that pose a risk to human health include those that can cause fire, explosion, annoyance-inducing illness, or even death. This poses a risk to human life and property. Everyone has a chance of coming into touch with dangerous vapors and gases. Due to the possibility of exposure to hazardous gases and vapors, utilizing gas sensors becomes inevitable. As a result, it is essential that they use a system or piece of equipment to identify these dangerous gases during their work to ensure the workplace is secure.

7.10.3 Infrared IoT Sensors

Optical or infrared sensors are objects that, in response to an incident, alter the resistance value or electrical conductivity. By reflecting light when it strikes an object, it may identify impediments. When an object passes through the front of the sensor, the sensor will detect it. There are infrared transmitters and receivers on infrared sensors [12]. The output will be the (white) infrared signal. The ordered infrared signal will also reflect back to the receiver (black) when an object is blocking it; this can be used to detect objects in front of it and to change the sensitivity of the detecting distance from close to distant.

7.10.3.1   Medical Care

Medical professionals measure skin temperature using infrared sensors. It is possible to measure the skin's average temperature using infrared thermography. The temperature of a person's skin can provide information about physiological issues with metabolism and thermoregulation. These sensors can provide critical information that can aid in clinical diagnosis and treatment evaluation.

7.10.3.2   Devices for the Home

In order to control motion and state activities, air conditioning and heating systems can employ infrared sensors to save energy and maintain a comfortable environment. To prepare and regulate temperature, the sensors can scan the space. They can calculate the heat from the window to modify cooling or heating requirements and show real-time critical environmental information.

7.10.4   Accelerometer Sensors

Another name for a gyroscope sensor [13] is an angular rate sensor or an angular velocity sensor. In three-axes directions, a gyroscope may detect the angular rate or angular velocity in degrees per second. It is known as a sensor utilized by navigational equipment. Our smartphone may be used to view 360-degree videos or images, and the gyroscope sensor allows the screen to spin when we move the phone's corner.

7.10.5   Gyroscope Sensors

Gyroscope sensors are used in smartphones and can automatically rotate the phone's screen. Today, the majority of cell phones use them. They improve user experience and offer details about their surroundings. Gyroscope sensors are necessary for 3-D games, like Pokémon Go, to display augmented reality (AR) because AR wouldn't function without them. Gyroscope sensors are utilized in robot systems, for camera shake detection in digital cameras, and electronic vehicle control systems. The optical image stabilization technology based on gyroscope sensors can be used to correct image blurring in cameras. By fastening the gyroscope sensor to a specific location, the gyroscope can be utilized to gather information about patient movement. When the data is processed by the application, the gyroscope displays the movement's direction in relation to time as a three-axis graph. The outcomes are seen in a 3-D style resembling the way motion is captured in a movie. The patient's movement characteristics are then contrasted with a healthy person's typical movements. Where the patient moves abnormally, it is simpler to diagnose.

7.11   EXAMPLES OF IoT AND SENSOR INTEGRATION IN A SMART CITY

At the moment, Europe is setting the standard for smart cities. Despite being the region of the globe with the highest level of urbanization, North America has trailed.

7.11.1   England

Milton Keynes, a sizable town in England's south, is a smart city that you should keep an eye out for. Three Smart Cities U.K. prizes in the categories of data, communications, and energy have been given to Milton Keynes. The development of the cutting-edge DataHub served as the foundation for this city's smart city initiative. This data center, which presently has more than 700 sets of data acquired for the development of urban resources, includes information on energy use, water use, mass transit, society, economy, and satellite data. It also allows the general public to have access to this information. Data hubs will be used in a variety of industries, such as mass transit, to connect data to users. One example is the MotionMap application, which shows the real-time movement of people and vehicles throughout the city and provides details on parking spaces, road routes, and traffic congestion estimates to assist city residents in making decisions.

7.11.2   Singapore

Despite having existed for more than 50 years, Singapore [14] does not have a plan to simply transform its city into a smart city. As an innovative nation, the country did develop a Smart Nation plan. The Singaporean government wants to use technology to increase employment and revenue. The use of the e-payment system to handle all commercial transactions will help the nation transition to a cash-free culture. Singapore places a strong emphasis on integrating technology into every part of life, whether it is gathering open government data for the public and private sector, or gaining access to public data to benefit from it. However, Singapore must pay attention to cybersecurity in data security by protecting the privacy of public information if they want to modernize their city with IT.

The public healthcare system is another area of concentration for Singapore's Smart Nation initiative. Singapore anticipates that its population will start to age in the next 15 years as a result of its attention to transportation and health issues. In order to handle this aging population, health services will need reliable transportation and enabling technologies. The elderly will be able to live comfortably because of this. In the field of transportation, Singapore has created a self-driving vehicle and TeleHealth, which is utilized to give patients access to medical systems even while they are at home.

The Singaporean government has also created a number of applications that aim to improve all facets of Singaporeans' quality of life, including health with the HealthHub app to record health data, travel with the MyTransport.SG app timetable route, safety with apps to help report incidents or accidents, and many other apps that the government has created.

7.11.3   Netherlands

Urban modernization initiatives are also carried out in a livable nation like the Netherlands [15]. One outstanding example is Amsterdam, which has been a smart city innovator since 2009 and has more than 170 projects that were created in conjunction with the public and commercial sectors. The intention is to find a solution to the gridlock issue, energy conservation, and creating circumstances for public safety.

Amsterdam wants to build a city that rotates the city's economy through the Circular Amsterdam project. Additionally, waste and pollution are reduced. Reusing resources is a critical part of sustainability, and there are many creative ways to do it.

CityZen, an initiative that has become synonymous with the city, calls for a complete transition to renewable energy from geothermal heat, biomass, wind, and solar energy. Additionally, it seeks to integrate this energy usage with urban infrastructure, construction, and an occupant's daily life, such as with solar energy storage strategies. Excess energy that is not used can potentially be sold.

Numerous projects are also being undertaken in fields including education, mobility and transit, infrastructure and technology, and citizen life, among others. They can monitor the specifics of many projects since they are a smart city, involving participants, operators, and follow-up through the city website as well.

7.12 SENSOR INTEGRATION FOR IoT

In a smart home system, temperature and humidity sensors are used to collect data on the temperature and humidity levels in different rooms of a house. This data is then transmitted wirelessly to a central hub that is connected to the internet and can be accessed remotely by the homeowner.

The integration of sensors in this system allows homeowners to monitor and control the temperature and humidity levels in their home remotely. For example, if the temperature in a room is too high, the system can automatically turn on the air conditioning to cool the room down. Similarly, if the humidity in a room is too high, the system can turn on a dehumidifier to reduce the humidity levels. The sensors used in this system are typically small and compact and can be easily installed in different rooms of the house. They are designed to be energy-efficient, with low power consumption to ensure that they can operate on battery power for long periods of time.

The data collected by the sensors is transmitted wirelessly to the central hub using a wireless communication protocol such as Zig-Bee or Wi-Fi. The central hub is responsible for collecting the data from the sensors and processing it, before transmitting it to a remote server or cloud-based platform for storage and analysis. To ensure the security and privacy of the data collected by the sensors, the system may incorporate encryption and authentication mechanisms to protect against unauthorized access or data breaches. Additionally, the system may include a user interface or mobile application that allows the homeowner to view and control the temperature and humidity levels in their home, as well as receive alerts and notifications if any abnormal conditions are detected.

7.13   FUTURE TRENDS FOR IoT IN SMART CITIES

7.13.1   Fire Alarm System

Sensors monitor the weather in open spaces and densely forested areas that could catch fire. Similar to this, sensors can identify building fires and alert nearby emergency services. Firefighters can plan ahead by using an IoT system's remote control and diagnostic capabilities to arrange personnel and vehicles. An IoT system notifies fire crews when a smoke alarm sounds, a seat finder transmits signals, or a water stream valve is turned on.

7.13.2   Systems for Inspecting Bridges

Bridge structural integrity is monitored by sensors that notify city engineers of any issues. Drones are utilized to inspect problematic hard-to-reach regions of bridges, and the engineer is quickly given the situation so they may use the smart bridge system software to solve the issue. The system conforms with nondestructive testing requirements and functions by mounting sensors to a structure, such as a bridge, to enable ongoing structural health monitoring. The sensors act as an early warning system that enables local governments and asset owners to schedule onsite inspections and maintenance; they are not intended to replace customary inspections.

7.13.3   Sensors for Waste Management

The finest method for cleaning the neighborhood with IoT smart technology is outlined in this trend. Sanitation personnel can clear waste out of the way of their routes thanks to IoT sensors that measure the amount of trash in the region. To help businesses save money and be more environmentally friendly, IoT fills level sensors, automates processes, and enhances waste management frameworks. IoT sensors are, in essence, a much better option for cities looking to assure linked, long-term growth.

7.14   IoT BENEFITS AND DRAWBACKS FOR SMART CITIES

The advantages of IoT in smart cities, the advantages and disadvantages of smart cities, and some of the most significant recent developments and difficulties in IoT-based smart cities will all be covered in this section.

The following are a few major advantages of IoT applications for smart cities.

Increased effectiveness and efficiency: The ability to access a vast amount of priceless information is one of the main benefits of IoT for smart cities. City officials can improve the city and the quality of life for its residents by making more effective judgements by appropriately assessing this information and data. IoT and cloud-based big DA for smart future cities can help cities more effectively detect high-risk locations and dispatch police. Similar to how citizens' wants and interests can be better identified and met by city officials.

Lower crime: By lowering crime rates, IoT-based smart city technologies make cities safer for their inhabitants. Police officers can work more effectively when using IoT technology for smart cities, such as internet-linked body cams and connected crime scenes. In fact, several communities are already implementing clever methods to build a safer neighborhood.

Improved environment: The amount of greenhouse gases in the environment is increasing. The adverse effects on the environment are, however, lessened by smart cities and energy solutions, such as renewable energy sources and energy-efficient structures. This raises the allure of a smart city as a technology. By enhancing air quality, an IoT-based smart city also lessens environmental pollutants. For instance, by putting IoT air quality sensors all across the city, cities can gather data on sources of pollution or pinpoint terrible air quality peak hours. Then, decision makers can keep an eye on the city's most polluted neighborhoods, redirect traffic to different ones, or choose for low-emission modes of transportation.

Superior services: Other significant advantages of a smart city include better healthcare and transportation options for residents. Intelligent traffic lights, connected cars, and smart parking all contribute to better transportation management. A better healthcare system is also aided by smart healthcare technologies like remote monitoring.

Less congestion in the traffic: Developing IoT and cloud-based apps for smart cities also aids in easing traffic congestion. City officials have the ability to divert heavy traffic by installing sensors in high-traffic locations. A smart parking system powered by IoT can also lessen traffic congestion.

7.15   DRAWBACKS OF IoT SOLUTIONS FOR SMART CITIES

Smart cities primarily benefit inhabitants in ways that are better for them. However, there are certain disadvantages because a smart city is built on technology.

- Privacy of citizens is less protected since city officials can access sophisticated systems and cameras to monitor city activity.
- Since IoT technology is the foundation of a smart city, a poor internet connection might impair city operations.
- Introducing smart technology into cities comes at a great expense. Therefore, it takes time for smart technologies to fully pay off.

7.16   IoT AND CLOUD COMPUTING IN SMART CITIES

IoT-based hardware generates a lot of data. As a result, they require a location to keep this data, but the storage, processing power, and performance of these devices are typically constrained. Cloud computing can help with this.

Shared resources like networks, storage, software, and servers are available in the cloud. The IoT sensor data that smart cities collect can be kept in the cloud's limitless storage. City officials can embed IoT data within IoT-connected electronic equipment thanks to cloud services and platforms for smart cities.

Public, private, and hybrid clouds are just a few of the current cloud service possibilities. Ridge, the most dispersed cloud in the world, maximizes cloud computing's potential. With no restrictions on latency or data residency, it enables enterprises to deploy apps anywhere, on any infrastructure.

7.17 CONCLUSION

The IoT has made sensors a crucial component of both our daily lives and city operations. So, to wrap up, what part does the IoT play in smart cities? IoT may gather, exchange, and analyze data, then provide the general public with solutions. With this, smart cities may effectively raise the social and economic conditions of its inhabitants. It can aid in data collection and system-based work process management. The growth of smart cities is expected to pick up steam over the next several years because of its practically infinite possibilities. However, the IoT will soon have a significant impact in a number of other areas as well.

In remote monitoring and control of environmental conditions, the sensors used in this system are designed to be small, energy-efficient, and wireless, with data transmitted to a central hub for processing and transmission to a remote server or cloud-based platform. The system may also incorporate security and privacy mechanisms to protect against unauthorized access or data breaches, as well as a user interface or mobile application for remote control and monitoring. With IoT, many firms are learning more about sensor technology and manufacturing processes.

References

[1] Hansain, S., D. Gaur, and V. K. Shukla, "Impact of Emerging Technologies on Future Mobility in Smart Cities by 2030," *2021 9th International Conference on Reliability, Infocom Technologies and Optimization (Trends and Future Directions) (ICRITO)*, Noida, India, 2021, pp. 1–8, doi: 10.1109/ICRITO51393.2021.9596095.

[2] Kuo, J. J., H. Huang, and K.-H. Chao, "An Intelligent Transportation System Framework Based on Internet of Things and Big Data Analytics," *J. Supercomput.*, Vol. 73, No. 9, 2017, pp. 3955–3968.

[3] Milasi, M. L., and R. Ciuciu, "The Role of Data Hubs in Smart Cities," *Int. J. Innov. Technol. Manag. Res.*, Vol. 5, No. 1, 2017, pp. 1–10.

[4] Zanella, A., N. Bui, A. Castellani, L. Vangelista, and M. Zorzi, "Internet of Things for Smart Cities" *IEEE Internet of Things Journal,* 2014, Vol. 1, No. 1, pp. 22–32, doi:10.1109/JIOT.2014.2306328.

[5] https://livejapan.com/en/in-tokyo/in-pref-tokyo/in-tokyo_suburbs/article-a0002533/.

[6] Adli, A. J. A., M. F. A. Aziz, and M. H. Marhaban, "A Review on the Applications of Data Analytics in Transportation Systems," *J. Telecommun. Electron. Comput. Eng.*, Vol. 9, No. 2–6, 2017, pp. 95–99.

[7] Kaur, N., "Smart City: A New Approach to Urban Development," *Int. J. Innov. Res. Sci. Eng. Technol.*, Vol. 5, No. 6, 2016, pp. 10316–10323.

[8] Rahman, M. M., S. I. Khan, and K. Salah, "Artificial Intelligence and Machine Learning for Internet of Things (IoT) Applications," *IEEE Internet of Things Journal*, Oct. 2020, Vol. 7, No. 10, pp. 9640–9657.

[9] Elnour, A. M. A., M. A. H. Akhir, and A. A. K. Abdullah, "A Review of Sensors Integration for Internet of Things Applications," *J. Telecommun. Electron. Comput. Eng.*, Vol. 9, No. 2–6, 2017, pp. 73–78.

[10] Ramasubramanian, K., and T. Mukherjee, "Wireless Sensor Networks: Current Trends and Future Directions," *Journal of Computer Science and Technology*, Vol. 31, No. 3, May 2016, pp. 509–536.

[11] Ali, S., S. A. Khayam, and T. L. Porta, "An Overview of Industrial Internet of Things (IIoT): A Roadmap to Future Industrial Productivity," *Journal of Industrial Information Integration*, Vol. 25, Sept. 2019, pp. 1–15.

[12] Sharma, R., P. Kumar, and R. K. Jain, "A Review of Energy Efficient Techniques for Wireless Sensor Networks," *Journal of Ambient Intelligence and Humanized Computing*, Vol. 9, No. 3, June 2018, pp. 865–889.

[13] Naqvi, H., M. A. Shah, F. A. Khan, M. S. Siddiqui, and M. U. Khan, "Sensors in Internet of Things (IoT): A Review," *Int. J. Comput. Appl.*, Vol. 168, No. 10, 2017, pp. 15–22.

[14] Chia, E. S., "Singapore's Smart Nation Program—Enablers and Challenges," *2016 11th System of Systems Engineering Conference (SoSE),* Kongsberg, Norway, 2016, pp. 1–5, doi: 10.1109/SYSOSE.2016.7542892.

[15] Harbers, M., and P. van Waart, "Valuing the Smart City: A Study of the Values of Different Stakeholders regarding Living Lab Scheveningen," In *Smart Cities in Smart Regions Conference Proceedings,* Meri Jalonen (ed.), Tampere, Finland, 2022, pp. 123–132.

8

IoT IN WIRELESS COMMUNICATION

8.1 INTRODUCTION

WSNs are one of the most crucial parts of IoT systems because they are primarily used to gather data from the environment and transmit it to the central controllers for additional processing. However, IoT sensors must be smarter than those in typical wireless sensor networks [1]. In particular, given their limited resources and the dynamic nature of the environment for the required IoT services, sensors in the IoT can not only perform typical functions, such as sensing information from the surrounding environment, but also make optimal decisions with minimal or no human intervention. Additionally, there are numerous difficulties in effectively controlling and maintaining the IoT's sensors due to the billions of devices connecting to the internet. As a result, new strategies that are more effective and adaptable to dynamic IoT networks must be created. Economic and pricing-based approaches have been widely used in IoT systems in addition to traditional approaches, such as optimization-based approaches.

The advantages of economic and pricing-based techniques over optimization-based ones include the following:

- Revenue generation is the main economic approach and most significant advantage. IoT systems need to generate as much profit as possible given their revenue and expenses.

- Since IoT components may belong to many organizations, such as sensor owners, spectrum providers, and data center operators, they may have a variety of goals and limitations. Pricing methodologies are offered to establish the best interactions between these self-interested and rational entities.

- To collect data from portable smart devices, IoT has used new sensing paradigms including crowdsensing networks and participatory sensing. Considering consistent scale of participants and to enhance the accuracy, coverage, and timeliness of the sensing results, incentive mechanisms using price and payment schemes can be used to entice users to contribute their data.

- Choosing the sensors with the highest remaining resources to carry out sensing activities is made possible by using economic and price models, such as auctions. This can ensure a trade-off between increasing network lifetime and giving sensors the necessary data quality. Furthermore, data redundancy can be quickly removed using price models without the need for complex calculations.

Even though there have been some studies on data collection and communication in wireless sensor networks, they have mainly concentrated on conventional methods. Additionally, there are studies on pricing strategies, such as [1–3], although they focus on problems with wireless or internet networks. Furthermore, the surveys on game theory applications for WSNs [4, 5] place more of an emphasis on generic issues than on issues explicitly related to pricing and economics. A survey that directly addresses the usage of economic models to cope with data gathering and communication in IoT devices is hard to find. A study in the current literature analysis on the economic models used in IoT networks is always helpful. Various economic models and pricing theories might be used to handle data gathering and communication problems in IoT systems through this study.

8.2   IoT DEFINITIONS IN WIRELESS

IoT is a broad notion, and several definitions are available. IoT, in general for wireless, refers to a self-configuring, adaptive, complex network that enables various devices and things, such as RFID tags, sensors, actuators, and mobile phones, to communicate and work together through specific addressing schemes in order to accomplish shared objectives. Along with the definition, several important IoT characteristics are listed as follows:

1. *Sense-making ability*: Things in an IoT environment can carry out sensing functions.

2. *Heterogeneity*: The IoT may accommodate a range of diverse communication devices, such as access point-based and peer-to-peer (P2P) style devices, as well as a variety of diverse underlying networks, such as wired, wireless, and cellular.

3. *Addressing modes*: IoT may enable transmissions in anycast, unicast, multicast, and broadcast modes.

4. *High reliability*: The IoT provides connectivity and trustworthy transmissions based on various technologies.

5. *Self-awareness*: The IoT has three self-capabilities:
   • High configuration autonomy;
   • Self-organization and self-adaptation to dynamic scenarios;
   • Self-processing of the enormous volumes of shared data.

6. *Secure environment*: The IoT provides the ability to withstand security concerns including network attacks (such as hacking and denial of service (DoS) attacks), authentication, data transfer confidentiality, data/device integrity, privacy, and trusted secure environment.

8.3   IoT ARCHITECTURE

Several IoT architectures have been offered [6–8] to satisfy the aforementioned criteria.

Figure 8.1 depicts an IoT architecture with the following several tiers:

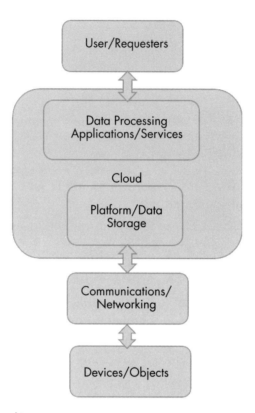

Figure 8.1 IoT architecture.

1. *Devices*: Because they have limited processing and storage capabilities, low-level devices like RFID tags, sensors, and smart devices perform only basic duties like collecting data from physical elements like the environment or keeping an eye on a specific region of interest. The devices can also connect to internet gateways to aggregate data or forward data, among other things.

2. *Networking and communications*: This layer consists of infrastructures for data communication and networking that transports information obtained from physical layer devices to higher layers, such as cloud platforms.

3. *Platform and data storage*: The purpose of this layer is to provide facilities for the access to and storage of data. It consists

of hardware and platforms in data centers or services in the cloud.

4. *Data management and processing*: This layer may consist of an application program that offers IoT consumers access services.

Resources and IoT services: Section resource management is a major problem to effectively offer IoT resources and services to customers since IoT is a heterogeneous large-scale system with a various resources and services.

The following resources and services can generally be sold in the IoT by using market and pricing models:

1. *Sensing information*: At the physical layer, sensors collect sensing information. Sensing data can be used to extract sensing information that can then be exchanged and priced to maximize owner profits.

2. *Power*: In order to function, IoT components like servers, access points, base stations, and sensors require energy. Energy suppliers can increase their revenue in monopoly markets by lowering the cost of the energy they supply to IoT components. Recently, self-energy recycling for IoT sensors through energy collaboration, in conjunction with the smart grid, has emerged as a practical alternative.

3. *Cloud services*: These are offered and exchanged by IoT users and include cloud data storage and computer services.

4. *Network bandwidth and spectrum*: In wireless networks, bandwidth and spectrum are valuable resources for data transmission. For instance, in cognitive radio networks, licensed and unlicensed users might dynamically trade spectrum in order to increase income and optimize spectrum use. Utilizing the sensing data and information stored in the cache, caching can be a practical method for conserving spectrum and network capacity in a large-scale IoT network, provided that the information value is not out-of-date and is temporally valid. Data and information services, such as information searching, data

mining, and data security, can be provided and integrated to enable IoT applications in providing services to IoT users.

5. *Location-based services*: The IoT offers location services to interested people, organizations, and the government by utilizing real-time geographic data from user devices like smartphones or tablets. The services include interior and outdoor location services like finding a person's or an event's location and learning where the closest eateries, coffee shops, and stores are.

The IoT's principal location-based services already generated a revenue of over £34.8 billion in 2020. The sensing data and information are the most crucial among these resources for IoT because they may optimize the utility and gain for owners and providers. An IoT business model was published in [9] to explain how IoT alters business paradigms. The value proposition of data resources serves as the model's central component, and the IoT business model essentially comprises of these four components. Setting the cost of the sensing data resources and promoting IoT users, or a customer's willingness to pay, are both parts of the value proposition. In IoT business models, selecting the right pricing strategy is crucial because the value proposition will generate the majority of the organization's revenues.

8.4   IoT AND WIRELESS COMMUNICATION COMPATIBILITY

What kind of communication would there be and how would the devices be connected? How will the protocols for wireless communication change?

Each IoT communication strategy is briefly described and illustrated in the sections that follow.

8.4.1   Satellite

Cell phone communication is possible at distances of 10 to 15 miles, thanks to satellite connections. Depending on the speed of connectivity, they go by the names GSM, GPRS, CDMA, GPRS, 2G / GSM, 3G, 4G / LTE, EDGE, and others. Because it enables devices like a phone to send and receive data through the cell network, this type of communication is most commonly referred to as M2M (machine-to-machine) in the context of the IoT.

Satellite Communication: Advantages and Drawbacks
Some pros of satellite communication are:

- Consistent connectivity;
- Constant compatibility.

Cons of satellite communication include:

- The device and smartphone cannot communicate directly; instead, a satellite connection is required;
- The monthly fee is high;
- Large energy consumption.

Utility meters that transfer data to a remote server, updated advertisements on digital billboards, or automobiles with internet connectivity are examples of satellite connectivity.

In the near future, when the cost of satellite communication is steadily declining, the usage of satellite technology may become much more viable and appealing for consumers. Satellite is useful for communication that consumes low data volumes, primarily for industrial purposes.

8.4.2 Wi-Fi

Wi-Fi is a WLAN that operates on the 2.4-GhZ UHF and 5-GhZ ISM frequencies and adheres to the IEEE 802.11 standard. Devices inside the range of Wi-Fi can access the internet (about 66 ft from the access point).

Pros and Cons of Wi-Fi
Pros of Wi-Fi include the following:

- Affordable;
- Universal smartphone compatibility;
- Well-controlled and protected.

Some cons of Wi-Fi are:

• Relatively high energy consumption;

• Erratic and inconsistent performance.

As an illustration of Wi-Fi connectivity, consider Dropcam transmitting live video via the community Wi-Fi rather than through an Ethernet LAN cable. Wi-Fi is helpful for many IoT connections, although these connections usually link to a remote cloud server rather than the smartphone directly. Due to its relatively high power consumption, it is also not advised for battery-powered devices.

8.4.3   Radio Waves

The simplest method of communication between devices is probably RF communications. Low-power RF radios implanted in or retrofitted onto electrical systems and devices are used by protocols like ZigBee and Z-Wave.

The range of Z-Wave is around 100 ft (30m). Its nation-specific RF band is employed. The 868.42-MHz short range device (SRD) band in Europe, the 900-MHz ISM or 908.42-MHz band in the United States, the 916-MHz band in Israel, the 919.82-MHz band in Hong Kong, the 921.42-MHz band in the regions of Australia/New Zealand, and the 865.2-MHz band in India are a few examples.

ZigBee is based on the IEEE 802.15.4 standard. The transmission distances are, however, restricted to a range of 10m to 100m because of its low power consumption.

RF: Advantages and Drawbacks

Pros of RF:

• Its technology is simple and uses little energy, thus it is independent of a smartphone's new features.

Cons of RF:

• Smartphones do not utilize RF technology, and without a central hub, the RF devices cannot be connected to the internet.

The usual television remote is an example of RF connectivity because it uses RF to let you change channels from a distance. Other examples include short-range, low-rate, wireless data transfer–requiring consumer and industrial devices such wireless light switches, electrical meters with in-home displays, traffic control systems, and more.

The RF communication protocol is helpful for large installations, like hotels, where a lot of devices need to be managed both locally and centrally. However, the technology might be superseded by Bluetooth mesh networks in the near future, as it becomes more and more antiquated.

8.4.4   RFID

The wireless identification of items by electromagnetic fields is known as RFID. Installing an active reader or reading tags that hold stored information, primarily authentication responses, is the norm. The experts refer to it as an active reader passive tag (ARPT) system. Long-range RFID can reach up to 200m, while short-range RFID is only approximately 10 cm.

Active tags are awakened by an interrogator signal from an active reader in an active reader active tag (ARAT) system. RFID operates on the following frequencies: 865 to 868 MHz (Europe), 902 to 928 MHz (North America), 3.56 MHz (10 cm to 1m), 120 to 150 kHz (10 cm), and 433 MHz (1m to 100m or 1m to 12m).

RFID: Pros and Cons

Benefits of RFID include:

- Does not require electricity;
- Reliable and generally accepted technology.

Some cons of RFID are:

- Expensive per card;
- Need to pass out tags as identification;
- Incompatible with smartphones.

Animal identification, factory data collecting, road tolls, and building access are a few examples. Inventory is also given an RFID tag so that the assembly line can follow the production and manufacturing status of that inventory. Pharmaceuticals are a good example of something that may be followed across warehouses. It is predicted that near-field communication (NFC) technology in smartphones will soon supplant RFID technology.

8.4.5 Bluetooth

Bluetooth is a wireless technology standard used for short-range data transmission (using short-wavelength UHF radio waves in the ISM band from 2.4 to 2.485 GHz). These two Bluetooth technologies appear to be extremely similar since they use the same frequencies as Wi-Fi. They serve various purposes, though. The three different Bluetooth technology models that are frequently discussed are as follows:

- *Bluetooth Low Energy (BLE (also known as Bluetooth 4.0))*: Originally developed by Nokia and currently supported by all of the top operating systems, including iOS, Android, Windows Phone, Blackberry, OS X, Linux, and Windows 8, BLE uses quick, low-power operation while maintaining communication range.

- *iBeacon*: This is Apple's trademark for a streamlined Bluetooth-based communication mechanism. It is a Bluetooth 4.0 sender that sends a universally unique identifier (UUID) that your iPhone can detect. This makes the installation process easier than it would have otherwise been for many vendors. Additionally, iBeacons like estimote.com or other options are simple to use even for consumers who lack technological expertise. Though distinct from NFC in terms of technology, iBeacon technology can be conceptually likened to it.

- Numerous goods, including phones, tablets, media players, and robotics systems, include Bluetooth technology. The method is very helpful in low-bandwidth scenarios when sending data between two or more nearby devices. Bluetooth is frequently used to send byte data with handheld computers or sound data with telephones (i.e., using a Bluetooth headset or transferring files). The discovery and deployment of services between devices is made simpler by Bluetooth standards. All of the services that Bluetooth devices offer can be advertised. Compared to previous communication protocols, this makes using services simpler because it allows for greater automation in areas like security, network address, and permission setting.

8.4.5.1   Comparison Between Bluetooth and Wi-Fi

Bluetooth and Wi-Fi have several applications and uses that complement one another:

- With an asymmetrical client-server connection and an access point at the center, Wi-Fi offers all traffic routing through the access point.
- Functions well in situations where high speeds are necessary and some degree of client configuration is feasible, such as network connectivity through an access node.
- Bluetooth and Wi-Fi can both establish ad hoc connections, but Wi-Fi Direct was recently developed to add additional Bluetooth-like ad hoc features.
- Functions effectively in straightforward applications requiring just two devices to connect, such as headsets and remote controls.
- There are Bluetooth access points, but they are not widely used.

Any Bluetooth gadget operating in discoverable mode instantly transmits the following data: name of the device, device class, services list, and technical information (i.e., device features, manufacturer, Bluetooth specification used, and clock offset).

Bluetooth's Advantages and Disadvantages

Bluetooth is a feature of every smartphone, and it is a technology that is constantly being developed and upgraded with new hardware.

Cons of Bluetooth include that the battery-powered iBeacon has a lifespan of one to two years and needs to be replaced frequently. If Bluetooth is turned off, there are problems with usage. Hardware capabilities vary quickly and require replacement.

Bluetooth technology is mostly used in the home entertainment, fitness, security, and healthcare sectors.

The hottest technology today is undoubtedly Bluetooth, yet its capabilities are frequently exaggerated or misinterpreted. Since different phones respond to Bluetooth in different ways, you will need to delve deeply into configuration and other options if the program becomes more than just for fun.

8.4.6 NFC

NFC [10] uses electromagnetic induction between two loop antennas placed in close proximity to one another, thereby forming an air-core transformer. It transmits data at speeds ranging from 106 kbit/s to 424 kbit/s over an air interface using the ISO/IEC 18000-3 standard in the 13.56-MHz ISM band, which is freely accessible worldwide. In NFC, an initiator and a target are involved; the initiator actively creates an RF field that can power a passive target (a chip that isn't powered and is referred to as a tag). As a result, NFC targets can have very basic form factors like tags, stickers, keychains, or card readers. If both devices are powered, NFC peer-to-peer communication is feasible.

Two modes exist:

- *Passive communication mode*: The target device responds by modifying the carrier field provided by the initiator device. In this mode, the target device can obtain its working power from the electromagnetic field supplied by the initiator, converting it into a transponder.

- *Active communication mode*: Both the initiator and the destination device generate their own fields in turn to exchange information. While awaiting data, a device turns off its RF field. Both devices normally have power supply in this mode.

NFC's Benefits and Drawbacks

Some benefits of NFC are:

- Provides a slow connection with a very easy setup;
- May be used to launch more powerful wireless connections;
- NFC is more suitable than older, less private RFID systems because it has a short range and offers encryption.

Cons of NFC include:

- Short-range may not always be practical because it is currently only supported by new Android phones and new iPhones with Apple Pay.

8.4.6.1   Comparison Between NFC and BLE

Mobile phones incorporate both NFC and BLE, two short-range communication technologies.

As seen in Table 8.1, the best communication protocol will ultimately depend on one's objectives and use-case. There will be a huge increase in the number of standard providers, particularly for mesh-networked technologies like iBeacons and GoTenna.

Table 8.1
Difference Between NFC and BLE

Features	NFC	BLE
Transfer speed	Slower (maximum data transfer rate: 2.1 Mbits/s)	Faster (maximum data transfer rate: 424 kbits/s)
Power consumption	Uses less power	Higher power consumption
Pairing requirement	No pairing required	Pairing required
Setup time	Quick setup time	Setup time may vary
Connection establishment	Automatically established connection	Connection establishment required
Range	Small range (around 20 cm)	Greater range
Compatibility	Compatible with ISO/IEC 18000-3 passive RFID infrastructures	Not compatible with RFID infrastructures
Energy protocol	Uses relatively less electricity	May consume more energy
Device power requirement	Can be used with unpowered devices	Requires powered devices
Mobile payment	Enables mobile payment	Can be used for mobile payment
Usage in smartphones	Primarily used for contact-only applications	Widely used for various applications
Factors for choosing protocol	Depends on objectives and use-case	Depends on specific requirements and context
Standard providers	Growing number of standard providers	Growing number of standard providers

8.5 IoT COMMUNICATION AND NETWORKING PROTOCOL DEVELOPMENT CHALLENGES

8.5.1 Software Definability

Since its conception, software-defined networking (SDN) has gained widespread acceptance in the traditional internet realm as a means of streamlining network management and configuration. Software-defined sensor networking (SDSN) [11] is a new idea that is realized by integrating SDN technology or concept into WSNs. Due to the separation of the control plane and the data plane, SDSNs allow for the software-defined definition of both the sensing and computing behaviors in addition to the networking behaviors. Thus, complete system management and orchestration are made possible by the ability to design the entire IoT infrastructure over the air, including sensing, transmission, storage, and computation. Additionally, it raises certain fresh issues that must be resolved.

For instance, conventional network-oriented SDN is controlled centrally. One controller might not be sufficient in the IoT situation, though, due to its widespread dissemination. Multiple controllers should be used in this situation. It is important to clearly specify the horizontal and vertical interactions between these controllers and the IoT devices (containing not only the sensors but also all the infrastructure-related devices). Furthermore, since it is well-known that standard SDN uses a lot of resources due to the addition of control overhead, resource-saving protocols are anticipated to achieve the lightweight software-definability of IoT. Security is another crucial concern with software definability. The infrastructure becomes more vulnerable while also becoming more elastic, flexible, and open thanks to the software-definability. The literature survey reveals that the protocols need to be strong and secure enough to handle any potential threats or weaknesses.

8.5.2 Integration, Compatibility, and Interoperability

As was previously indicated, there are already many different communication protocols available. When the devices must interact with one another, this creates an interoperability issue. It is necessary for items to be found, accessed, controlled, and semantically linked to one another in order to realize the full potential of the IoT, according to Blackstock and Lea [12]. A greater level of interoperability is required to facilitate this interaction. Utilizing web technology, which has evolved from the IoT to the Web of Things (WoT) [13] suggests, is a successful way to increase interoperability. In support of this idea, other protocols (such as 6LoWPAN) have also been developed. On several subjects, we have still not managed to come to a consensus. For instance, there is still no consensus on how to expose physical objects online. Additionally, several low-power networking protocols (such as ZigBee, Z-Wave, and Bluetooth) are created for applications that are specialized to a given domain and have special features. Standardization at the hardware level is needed to address interoperability problems at this level. A global standard is therefore urgently needed to address the entire IoT interoperability challenge.

It is also greatly desired that the IoT can be easily connected with some other recently emerging technologies, in addition to the interoperability between devices (e.g., cloud, or Blockchain). Additionally, this creates some fresh integration issues. As an illustration, the IoT ecosystem's reliance on centralized cloud infrastructure and the absence of security procedures may result in numerous cybersecurity threats. There are still several difficulties with the integration of IoT and Blockchain, including issues with data security, privacy protection, access control, and resource management.

As the IoT develops, a variety of IoT architectures and protocols emerge, which raises the issue of compatibility. We cannot abruptly alter all the protocols on all the devices at once, just as IPv4 evolved into IPv6, for example. The newly established protocols would therefore be more compatible with the existing protocols from the perspective of interoperability. However, the majority of currently available solutions limit compatible devices to a specific protocol. So, interoperability presents yet another hurdle in the way of IoT development.

8.5.3 Network Computing

It is anticipated that billions of IoT devices will be connected to the internet as it develops. The IoT devices on the premises won't be able to adequately process the bulk of the data that will be produced. Although it is possible to transfer this data to the cloud, doing so easily and sending the raw data may strain the network and use a lot of energy. The growth of IoT will not be aided by this. In order to achieve this, Zeng et al. [14] made a compromised solution known as in-network computing (INC) or computing in the network (COIN), which allows all networked devices to process data along the transmission path.

Traditionally, computers have been kept out of the design of communication and networking protocols like transmission control protocol (TCP's) congestion control, flow control, and different routing algorithms. Therefore, the entire protocol stack needs to be refactored with the introduction of INC. For instance, the packet processing semantics must be taken into account by congestion control because even the identical routine may have varying end-to-end delays due to the various processing semantics. Given that routing devices (other than routers) are capable of handling some data processing tasks, the routing protocols should also take computational power into account. The distinction between networking and computing is blurred by INC, in which any device may take part. The IoT device may also be a routing device, and some routers along the routing path may process the IoT data. As a result, the protocols need to be able to handle the infrastructure's high degree of heterogeneity and effectively adapt to various devices with various capabilities.

8.5.4   For and by AI

The INC combines computers and networking. In order to increase production and efficiency, such convergence both makes possible and necessitates a high level of automation and intelligence. In order to achieve this objective, AI is emerging as a critical enabling technology. Reinforcement learning, a type of AI technology, has proven to have great potential for route management and congestion reduction in conventional networks. It is without a doubt applicable to managing networking and communication for the IoT, but it still has a lot of problems. First off, there is a serious adaptation issue with AI technology. The agent who has been trained for one scenario may struggle in another. Realizing a comprehensive solution is still challenging. Asynchrony between these agents, which is used by various service providers to meet various needs, may also result in performance degradation, according to Sheth et al. [15]. Furthermore, the AI-based solution might be computationally expensive, may require a lot of time and resources to train, and might even be too expensive for particular IoT devices. Also, an intelligent agent needs a lot of data to function properly. Additionally, the gathering of this data adds some extra overhead, which even competes with data transmission. When the information gathered by IoT devices is used for automatic or semi-automatic control applications, latency constraints are very important as put forth by Kaminski et al. [16]. Therefore, using AI technology to control IoT communication and networking appears to be a potential approach, but more work is still needed to address these issues.

Numerous IoT applications now widely use AI technology to process IoT data. The extremely high computing power consumption of AI technologies is well-known. Some IoT devices may not be able to meet the criterion. There are two basic methods for handling this issue. The AI technology themselves are one strategy. To reduce the amount of processing required, we might, for instance, apply pruning and compression to particular deep neural network (DNN) models. Another strategy focuses on task scheduling. For instance, we might divide a sizable DNN model into a number of interdependent jobs that are then delegated to networked devices for distributed processing. Actually, we may attempt combining the two strategies to come up with a better answer that will better match the supply and demand for processing power and, ultimately, the high quality-of-service (QoS) of AI-based IoT applications. However, there are still many unique solutions needed for this enormous task.

8.5.5   Energy Efficiency

As many IoT devices are powered by capacity-limited batteries and communicate with one another wirelessly, energy efficiency has already been intensively investigated in the IoT community. Applying the right communication technology is crucial for maintaining connection linkages and supporting real-time transmission in an energy-efficient way, according to Ramamurthy and Jain in [17]. As a result, low-power communication technologies like Bluetooth, ZigBee, and long-range (LoRa) that are appropriate for the IoT, have been continuously developed. In the meantime, numerous energy-efficient routing strategies have been put forth by the research community to increase network lifetime as proposed by Gopika and Panjanathan [18] and cut down on energy use. Routing protocol for low-power and lossy networks (RPL), 6LoWPAN, and ZigBee IP are a few examples of low-power network encapsulation protocols that are constantly developing. However, the recently adopted information-centric networking (ICN) necessitates an IoT device capable of more than just detecting and communicating but also to do some computational tasks.

These factors are related to one another, thus focusing exclusively on energy-efficient communications does not seem like the best option. It is strongly anticipated that new protocols will be developed that can balance the energy requirements of sensing, transmission, and processing.

Exploiting renewable energy from the environment to reach the aim of zero-carbon IoT is another trend in development to address the issue of energy efficiency. Technology for capturing energy has also advanced quickly in recent years. The idea of incorporating these technologies to supply renewable energy to IoT devices is enticing. Although numerous methods have been put forth as of late for effectively utilizing renewable green energy to increase the lifespan of IoT devices or decrease the use of brown energy, this movement is still in its early stages. In addition to working independently, communication and networking protocols should be built with energy considerations. New energy-efficient communication technologies have also recently drawn a lot of interest. For instance, researchers recently supported backscatter communications for the IoT as explained by Zhang et al. in [19]. The rise and adoption of these new technologies also call for related communication and networking protocols.

8.5.6 Multiple Communication Channels

The majority of IoT networking and communication protocols focus on electromagnetic or optical communications. However, some unique situations (such as those found underground and underwater) need for specific protocols and other medium-oriented communication methods. For instance, Khalil et al. [20] claim that the underwater IoT is one of the revolutionary technologies for both ocean research and biodiversity preservation. In the harsh undersea environment, electromagnetic communication performance drastically decreases. Acoustic communication is frequently recommended for underwater IoT to address this issue. Similar to this, IoT devices buried underground are typically covered with a layer of asphalt and soil. The rate of electromagnetic waves, and thus the effectiveness of communication, will be significantly impacted by soil texture (e.g., pore spaces, clay, sand, and silt particles) as explained by Vuran et al. in [21]. Therefore, it is crucial to comprehend the effects of various layers of communication medium on the signal propagation and develop communication and networking protocols in a way that takes environmental factors into account as suggested by Salam and Raza in [22]. The realization of a globally omnipresent IoT system is the ultimate objective of IoT development, and this necessitates that all IoT devices be seamlessly networked and allow unrestricted communication, regardless of their location or preferred method of communication. Another interconnection difficulty relating to how to adjust to these many communication mediums and technologies is raised by this. To increase their adaptability, protocols should be designed using a cross-layer approach.

8.5.7   Privacy and Security

This is another issue that is frequently broached. IoT devices with limited capacity are readily compromised by malicious hackers who then launch assaults against the entire network. Malicious actions could also compromise the IoT data's integrity and authentication. DoS attacks, for instance, can overwhelm the network and bring down communication between the devices and their source. An IoT device could be programmed to inject a bogus overflow indicator in order to stop a legitimate procedure. Private information could be revealed if unscrupulous attackers overheard the conversation. In order to contaminate the system and cause it to behave incorrectly or even dangerously, the attackers can also inject bogus data. Many IoT gadgets and systems, such as smart cities and automated driving systems, are gravely threatened by these attacks. Despite the various remedies that have already been suggested, new threats are constantly arising. Along with pervasiveness, the openness of IoT is another development that has the potential to maximize its power. Openness, meanwhile, always brings with it security risks. In particular, the advent of software-definability made it possible to control IoT devices remotely. Although this increases the IoT's flexibility and openness, it also raises the risk of hostile attackers compromising either the controller or a device, which increases security risks. Additionally, this new trend calls for greater security remedies.

8.5.8 Mobility in IoT Devices

Numerous heterogeneous communication devices that are often re-source-constrained and in need of effective routing protocols to enable data transmission from source to destination may be present in an IoT system. Mobile IoT devices have become more popular in recent years as an alternative to installing fixed dedicated IoT devices to improve flexibility or increase coverage. Examples of IoT devices include UAVs and unmanned ground vehicles (UGVs) that are fitted with various sensors, such as cameras, and can be used to develop crowd sensing applications by enlisting a large number of mobile handsets. Clearly, in contrast to standard IoT infrastructure, another problem in this situation relates to mobility. The management of mobility has received some recent attention. For instance, by leveraging hierarchical IPv6 address allocation to manage mobile devices, Santos et al. in [23] propose Mobility Matrix (Matrix) as an alternative to conventional routing protocols for the IoT.

Sometimes it is challenging to ensure the network connection's QoS when moving. While some IoT applications require dependable and low-latency data connection, there could be a significant delay and packet loss in the transit. Numerous creative solutions to the mobility problem have been put forth during the last few decades. To provide easy and efficient backward compatibility with standard protocols, Hossein et al. [24] and Fotouhi et al. [25] included an active handover mechanism (referred to as smart-HOP) in RPL. Despite this, mobility continues to represent a significant problem to the management and orchestration of the IoT, particularly with the introduction of new telecommunications technologies like 5G, B5G, and 6G. All of these technologies have several connections, and some even incorporate space-air-ground-sea networking. In addition to the interconnection issue stated previously, the mobility in the 3-D environment offers a significant obstacle to IoT networking and communication.

8.6 CONCLUSION

The difficulties that may arise when designing and implementing IoT communication and networking were discussed in this chapter while taking into account recently developing technologies, concepts, and trends in IoT communication, networking, and computing. It is important to remember that these difficulties are interrelated. Some of them are related to one another and will influence future IoT systems. Researchers and engineers should put more effort into these areas to enhance the development of IoT.

References

[1] He, H., K. Xu, and Y. Liu, "Internet Resource Pricing Models, Mechanisms, and Methods," *Networking Science*, Vol. 1, No. 1–4, 2012, pp. 48–66.

[2] DaSilva, L. A., "Pricing for Qos-Enabled Networks: A Survey," *IEEE Communications Surveys & Tutorials*, Vol. 3, No. 2, 2000, pp. 2–8.

[3] Gizelis, C., et al., "A Survey of Pricing Schemes in Wireless Networks," *IEEE Communications Surveys & Tutorials*, Vol. 13, No. 1, 2011, pp. 126–145.

[4] Shi, H.-Y., W.-L. Wang, N.-M. Kwok, and S.-Y. Chen, "Game Theory for Wireless Sensor Networks: A Survey," *Sensors*, Vol. 12, No. 7, 2012, pp. 9055–9097.

[5] Machado, R., and S. Tekinay, "A Survey of Game-Theoretic Approaches in Wireless Sensor Networks," *Computer Networks*, Vol. 52, No. 16, 2008, pp. 3047–3061.

[6] Atzori, L., A. Iera, and G. Morabito, "The Internet of Things: A Survey," *Computer Networks*, Vol. 54, No. 15, 2010, pp. 2787–2805.

[7] Uckelmann, D., M. Harrison, and F. Michahelles, "An Architectural Approach Towards the Future Internet of Things," in *Architecting the Internet of Things*, Germany: Springer, 2011, pp. 1–24.

[8] Gluhak, A., S. Krco, M. Nati, D. Pfisterer, N. Mitton, and T. Razafindralambo, "A Survey on Facilities for Experimental Internet of Things Research," *IEEE Communications Magazine*, Vol. 49, No. 11, 2011, pp. 58–67.

[9] Niyato, D., D. T. Hoang, N. C. Luong, P. Wang, D. I. Kim, and Z. Han, "Smart Data Pricing Models for Internet-of-Things (IoT): A Bundling Strategy Approach," Vol. 30, No. 2, 2016, pp. 18–25.

[10] Cao, Z., et al., "Near-Field Communication Sensors," *Sensors*, Vol. 19, No. 18, Sept. 2019, p. 3947.

[11] Puente Fernández, J., L. García Villalba, and T.-H. Kim, "Software Defined Networks in Wireless Sensor Architectures," *Entropy*, Vol. 20, No. 4, March 2018, p. 225.

[12] M. Blackstock, and R. Lea, "IoT interoperability: A Hub-Based Approach," in *2014 International Conference on the Internet of Things (IoT)*, 2014, pp. 79–84.

[13] Kumar, S., P. Tiwari, and M. Zymbler, "Internet of Things is a Revolutionary Approach for Future Technology Enhancement: A Review," *J Big Data*, Vol. 6, No. 111, 2019.

[14] Zeng, L., J. Wang, Y. Zhang, and X. Zhang, "In-Network Computing for the Internet of Things: A Survey," *IEEE Communications Surveys & Tutorials*, Vol. 23, No. 2, Marc 2021, pp. 1026–1059.

[15] Sheth, J. N., R. Sisodia, and A. Sharma, *The Future of Marketing: A New Approach to Connecting with Consumers,* United Kingdom: Routledge, 2020.

[16] Kaminski, J., T. Callaghan, S. Marshall-Pescini, and B. Hare, "Dogs Follow the Gaze of Humans to Obtain Information About Novel Objects," *Scientific Reports,* Vol. 7, No. 1, 2017, p. 12914.

[17] Ramamurthy, A., and P. Jain, "The Internet of Things in the Power Sector: Opportunities in Asia," *Asian Development Bank,* Mandaluyong, Philippines, 2017.

[18] Gopika, D., and R. Panjanathan, "A Comprehensive Study on Various Energy Conservation Mechanisms in Wireless Sensor Networks," *IEEE International Conference on Advances in Computing, Communication, and Intelligent Systems (ICACCI),* 2020, pp. 1–6.

[19] Zhang, Y., J. Chen, L. Zhang, and L. Chen, "Backscatter Communication for Internet of Things: A Survey," *IEEE Communications Surveys & Tutorials,* Vol. 21, No. 4, 2019, pp. 2975–3002.

[20] Khalil, M. A., M. Elhoseny, A. E. Hassanien, and A. K. Sangaiah, "Underwater Internet of Things (uIoT): A Survey on Technologies, Challenges, and Applications," *IEEE Access,* Vol. 5, 2017, pp. 18940–18963.

[21] Vuran, M., M. Yaghoobi, and F. Saygin, "Electromagnetic Wave Propagation in Soil: A Review," *Progress in Electromagnetics Research B,* Vol. 72, 2018, pp. 1–34.

[22] Salam, A., and U. Raza, "A Survey on Underwater Wireless Sensor Networks: Challenges and Opportunities," *IEEE Access,* Vol. 8, 2020, pp. 127213–127230.

[23] Santos, I., P. Costa, and E. Monteiro, "Mobility Matrix: A Hierarchical Routing Protocol for the Internet of Things," *IEEE Internet of Things Journal,* Vol. 5, No. 3, 2018, pp. 1868–1877.

[24] Hossein, M., D. Moreira, and M. Alves, M., "mRPL: Boosting Mobility in the Internet of Things," Ad Hoc Networks, Vol. 26, 2015, pp. 17–35.

[25] Fotouhi, H., M. Zuniga, and M. Alves, "Smart-HOP: A Novel Mobility Management Scheme for RPL," *Proceedings of the 14th International Conference on Green, Pervasive, and Cloud Computing,* 2015, pp. 73–82.

9

IoT IN SURVEILLANCE AND SECURITY

9.1   INTRODUCTION

By 2025, it is anticipated that the IoT market will have grown to $1.5 trillion. Isn't that number mind-boggling? To put that in context, the IoT market was estimated to be worth $250 billion in 2019. This is a 600% growth. By 2025, there will likely be 25 billion connected devices, with smartphones accounting for 24% of those, or 6 billion gadgets. Industry 4.0 is the largest winner from the IoT revolution. Only if we can safeguard the vast amounts of data streaming through billions of IoT connections will the IoT revolution truly take off. This leads to our main subject of interest: IoT security.

IoT security entails protecting networks, software, hardware, and other elements that handle, receive, store, and process data. Working with sensors for everything from your garage door rolling up when your car reaches the perimeter of your property to your lights going on the instant you walk into a room is the new normal. These sensors gather information and transmit it to a command center that processes it and responds. This information must be protected. You must secure your devices. It is necessary to secure the entire network. IoT security deals with issues like these. It includes approaches, strategies, and technologies to protect you from hackers who repeatedly return to your networks in search of weaknesses.

IoT security is cloud-based, as opposed to device-based security, such as that of a laptop or smartphone. It is the cornerstone of the eco-system of IoT, big data, and cloud computing. IoT devices assist with data generation and collection, while the big data platform handles analytics. The cloud computing system handles additional facets of data mobility as well as data storage and processing. Everything that happens in the cloud and on IoT devices is well protected, thanks to IoT security.

Major actors in technology, industries, logistics, commerce, and governmental organizations believe that IoT security, not merely IoT device interconnectivity, will determine how successful Industry 4.0 will be. The current scenario supports the implementation of a strong IoT security management plan to address IoT security concerns. The best strategy to stop and address IoT security risks and vulnerabilities at the design level is to implement security-by-design architecture.

How would it feel to be unable to handle a 100-ton machine on a construction site? What would happen if the signaling system for the subway failed? Even imagining it is challenging. IoT devices connect complicated systems and devices, building complex networks, in-cluding power distribution, water management, traffic management, smart homes, and a ton of other systems and devices. Although IoT connectivity has many benefits, the worst is yet to come. Any breach in the security of these intricate networks could result in a disastrous situation. The threat might be on a national scale if a military net-work, nuclear facility network, or power transmission infrastructure is compromised. For the intricate and delicate networks that keep this world turning, robust IoT security is essential. A cybersecurity team faces a variety of IoT security difficulties, according to cybersecurity experts.

Let's use an intelligent vehicle manufacturing factory as an ex-ample. The factory's efficiency and competency serve as a showcase for what the IoT revolution may eventually bring. On the other hand, the same facility provides a compelling argument for appreciating the significance of IoT security. Hackers can gain access to crucial systems and operations running on the factory floor by breaching the factory's network. They will eventually gain control of the *privilege escalation* permissions. Hackers might potentially endanger hundreds of lives by messing with the settings of a manufacturing or assembly unit in the car industry. Similar to this, there is a possibility for more human loss if they are able to access a medical command control that oversees hundreds of medical equipment like artificial pacemakers.

Even household appliances like smart TVs, refrigerators, and closed-circuit televisions (CCTVs) pose a security risk. These devices are set up to connect to the household networks using their default passwords. This leaves the entire network vulnerable to malicious attackers. It's crucial to safeguard a network's endpoints all together. Data transfer between devices and the cloud should always be encrypted. This reduces the risk even in the event of a data breach.

9.2 THE IoT SECURITY FRAMEWORK: AN OVERVIEW

The majority of instructions based on different IoT security frameworks rigorously adhere to protocols and preestablished policies that are carried out through the cloud. Businesses adhere to a set of compliance standards as mandated by regional legislation, depending on the nature of the industry, the quantity and quality of data collected, data processing, and other factors. Even though this is true for safe data processing on IoT devices, manufacturers and customers need still be aware of their procedures. The IoT security framework mainly consists of four levels: device physical layer, end communication layer, core network processing layer, and the edge network (cloud). Table 9.1 provides an overview of the security concerns and mitigation strategies for each level of the IoT security framework. By understanding these security concerns, organizations can take steps to protect their IoT devices and networks from cyberattacks.

Security concerns and mitigation strategies for each level are discussed next.

9.2.1 Physical Layer

9.2.1.1 Design for Security

Strict implementation of security in IoT by design is necessary. Due to the fact that it is integrated into the system on chip (SoC), the development team should value the security feature on par with the device

Table 9.1

IoT Security Framework: A Table of Security Concerns and Mitigation Strategies

Level	Description	Security Concerns	Mitigation Strategies
Device physical layer	The physical layer is the hardware and firmware of the IoT device.	• Security-by-design is not always implemented. • User access credentials are not always kept private. • IoT devices are not always tamper-resistant.	• Implement security-by-design principles. • Use strong passwords and multifactor authentication. • Make IoT devices tamper-resistant.
Edge network	The edge network is the network that connects IoT devices to the cloud.	• The edge network is vulnerable to cyberattacks. • Edge devices are not always consistent in their security protocols.	• Use strong security protocols and authentication mechanisms. • Make sure that edge devices are consistent in their security protocols.
Core network	The core network is the network that connects the edge network to the cloud.	• The core network is a major entry point for cyberattacks. • Conventional cybersecurity techniques are not always effective against sophisticated cyberattacks.	• Use strong security protocols and authentication mechanisms. • Monitor the core network for signs of cyberattacks.
Processing layer (cloud)	The processing layer is the cloud where IoT data is processed and stored.	• The cloud is a major target for cyberattacks. • Cloud assets are not always secure.	• Use strong security protocols and authentication mechanisms. • Secure cloud assets.

itself. By doing this, IoT security risks are reduced for the duration of the IoT device. Only a secure mechanism should be used to distribute patches and firmware updates.

9.2.1.2 Using the Gadget

User access credentials must always be kept private and confidential. There should be action taken to stop abusive log-in attempts and brute force unlocking. IoT security concerns can be reduced by careful testing. Access to sensitive data should require multifactor authentication (MFA). MFA is a security process that requires users to provide two or more pieces of evidence to verify their identity before being granted access to a system or resource. This can include things like a username and password, a security code sent to a mobile device, or a fingerprint scan. MFA is an effective way to prevent unauthorized access to systems and data. It makes it much more difficult for attackers to gain access, even if they have stolen a user's password.

Here are some of the benefits of using MFA:

- It makes it much more difficult for attackers to gain access to systems and data.
- It can help to reduce the risk of data breaches.
- It can help to improve the security of your organization's systems and data.
- It can help to protect your organization from financial losses.

9.2.1.3   Antitamper and Detection Mechanisms

Manufacturers must make sure that the IoT device cannot be tampered with during shipping and installation using only a few tools. There should be a thorough detection mechanism built in to notify the command control if necessary. When purchasing IoT devices, users may choose better if they comply with specific security certifications. Due to the current increase in eavesdropping incidents, which are frequently carried out using low-cost IoT devices, this is of utmost relevance.

9.2.1.4   Keeping the Customer Informed

Users should always receive current, readable, and simple-to-understand information regarding the device. Users can secure their accounts by instantly changing the password or taking other preventive measures, especially during an incident where a data leak is very likely.

9.2.2   Edge Network

9.2.2.1   Interconnection of IoT Devices at the Communication Layer (Wired or Wireless)

Through a wired or wireless interface, the edge network enables communication between numerous IoT devices connected to the same network. It creates the necessary protocols for data processing and sharing among interconnected IoT devices. This line of communication is vulnerable to numerous cyberattacks.

9.2.2.2   *Edge Computing*

Edge computing is a distributed computing paradigm that brings computation and data storage closer to the edge of the network, where the data is generated. This can improve data processing performance and reduce latency, which can lead to more actionable results.

9.2.2.3   *Consistency Across All Devices*

A network's most common linked devices frequently come from multiple manufacturers. To reduce any hiccups during device connectivity to the network and to other devices, the majority of IoT device developers and manufacturers worldwide choose simple authentication and security protocols. As a result, threat actors can easily target them. Every device on the network as well as the network itself has been hacked, not only the device and the data it stores. Reducing IoT security vulnerabilities on such networks requires the establishment of a uniform security policy.

9.2.3   Core Network

9.2.3.1   *Connects IoT Devices to the Cloud*

The main line of communication between the cloud and the IoT device(s), enabling data mobility, is the core network. This entry point into a network is frequently targeted by hackers. In the wake of sophisticated and intricate cyberattacks, conventional cybersecurity techniques have failed. As a result of coordinated and sophisticated cyberattacks, the core network channel presents various IoT security difficulties.

9.2.3.2   *Security and Endpoints Duty Are Related*

The percentage of IoT security hazards linked with the network keeps growing as more IoT devices are connected to it. Even with the greatest IoT security measures in place, securing a network with hundreds of thousands of IoT devices requires ongoing network monitoring.

9.2.4   Processing Layer (Cloud)

9.2.4.1   *Processing and Big Data Analytics*

Processing and data analytics are done primarily in the cloud. This is by far the most important component need for an IoT device to perform as intended. Data is sent from the IoT device to the cloud for additional processing. One can comprehend long-term usage patterns and other associated user patterns after evaluating the data. Most cloud assets consist of applications, network management systems, and data storage.

9.2.4.2 *Putting in Place Development, Security, and Operations*

Development–security–operations is abbreviated DevSecOps. This is a step up from DevOps, which already exists and concentrates on development and operations. Adopting DevSecOps helps in incorporating greater security into the code and the device from the very beginning because of the agile development approach and a single large team working on projects. The programmers can use IDE security plug-ins, apply threat modeling guidelines, and follow secure code standards. These are tools that can be integrated into an IDE to help developers find and fix security vulnerabilities in their code. IDE security plug-ins can be used to scan code for known vulnerabilities, provide warnings about potential security risks, and help developers follow secure coding practices.

IDE security plug-ins can be a valuable tool for developers who want to improve the security of their code. However, it is important to note that IDE security plug-ins are not a silver bullet. They can help developers find and fix security vulnerabilities, but they cannot guarantee that code is completely secure. It is important to use IDE security plug-ins in conjunction with other security measures, such as code reviews and penetration testing.

Some of the benefits of using IDE security plug-ins are:

- Can help developers find and fix security vulnerabilities in their code early in the development process. This can help to prevent vulnerabilities from being exploited and can save time and money.
- Can help developers follow secure coding practices. This can help to reduce the risk of vulnerabilities being introduced into code in the first place.
- Can help developers to be more aware of security risks. This can help them to make better decisions about how to develop and secure their code.

Here are some of the drawbacks of using IDE security plug-ins:

- Expensive;

- Can be difficult to configure and use;
- Can produce false positives;
- Can only find known vulnerabilities.

Overall, IDE security plug-ins can be a valuable tool for developers who want to improve the security of their code. However, it is important to note that IDE security plug-ins are not a silver bullet. They should be used in conjunction with other security measures to ensure the security of code.

9.2.4.3   A Variety of Endpoints

It might be challenging to secure various endpoints that connect to the cloud from different IoT devices. Because not all connections are equally secure, it is necessary to constantly check for the existence of malware and other bots.

9.3   THREATS TO IoT SECURITY AND SOLUTIONS

While IoT devices offer many benefits, they also pose a number of security risks. Some of the potential security breaches in IoT surveillance and security include:

- *Data breaches:* IoT devices can collect a lot of sensitive data, such as video footage, audio recordings, and location data. If this data is not properly secured, it could be hacked and stolen by criminals.
- *Malware attacks:* IoT devices are often vulnerable to malware attacks. Malware can be used to take control of devices, steal data, or disrupt operations.
- *DDoS (distributed denial-of-service) attacks:* IoT devices can be used to launch DDoS attacks. DDoS attacks are designed to overwhelm a website or network with traffic, making it unavailable to users.
- *Physical attacks:* IoT devices can be physically attacked to steal or damage them. This could allow criminals to gain access to sensitive data or disrupt surveillance and security operations.

In addition to the security risks listed, there are a number of other potential security issues that could arise from the use of IoT devices in surveillance and security. For example, IoT devices could be used to track people's movements without their knowledge or consent. This could raise privacy concerns, particularly in sensitive areas such as schools and hospitals. Additionally, IoT devices could be used to create surveillance networks that are difficult to monitor or control. This could pose a threat to civil liberties and could be used to target or oppress certain groups of people.

Sections 9.3.1 to 9.3.5 list IoT security risks which have been put up by experts, along with information on how businesses may stay safe from them in the present. It is advised to always adopt a zero-trust policy when it comes to transferring log-in information and other data between devices.

9.3.1 Botnet-Based DDoS Attacks

One ant might not have much strength. But even the mightiest of men can be defeated by an ant colony. Hackers make the most of this information since they are certain of it. The computing power of one IoT device is little. However, the power these IoT gadgets can produce when combined with thousands of them is ridiculous. Hackers infiltrate IoT devices with malware, which replicates the attack and infects further IoT devices. These tools are then employed to flood the targeted server with requests, resulting in its failure.

Nobody can reliably distinguish between legitimate traffic and (DDoS) attack traffic in real time. Although a DDoS assault cannot be stopped, it can be mitigated by using solutions like web application firewall (WAF). By serving as a reverse proxy, it shields the targeted server from harmful traffic. Other options include blackhole routing, which sends all traffic into a black hole and removes it from the system, and rate limiting, which restricts the number of requests a server can receive. Another method of launching DDoS assaults is IP spoofing. The use of ingress filtering can help to limit this. Limiting hardware resources and bandwidth is a smart strategy to stop DDoS attacks. Botnets are employed to send copious amounts of traffic to the server or device, rendering it unusable. Internet service company Dyn was the target of a significant DDoS attack in 2016. This caused a serious outage.

9.3.2   Zero-Day Vulnerabilities

In several instances, hackers discovered vulnerabilities in program code before developers could. Hackers take advantage of the situation and utilize these exploits to break into networks and cause havoc. The hacking organization Elderwood's "Operation Aurora" is a prime example of a zero-day exploit. The victims included numerous tech behemoths including Google, Yahoo, Adobe Inc., and dozens of other businesses. According to reports, sensitive information was compromised. Zero-day exploits can be prevented through avoiding open-source code, thorough code testing, beta testing, and code review by professionals outside the organization. The development team should look for ways to infiltrate the application while adopting a hacker's perspective if the code has already been pushed. Exploits can be found using regular vulnerability scanning, input validation, sanitization, WAF deployment, and patch management.

9.3.3   Ransomware Assault

More than 90 countries have gross domestic products (GDPs) that are less than the $20 billion that ransomware assaults are expected to cost in 2022. The C-suite and security managers may worry about the security of their digital assets as a result of this statistic. Critical data, IoT devices, security systems, thermostats, and everything else that they have access to are taken over by hackers, who then demand a ransom. They might give up control of the assets after receiving the cash, or they might not.

Businesses should frequently have a backup of their important data. This is useful when they must choose whether to pay a ransom to a hacker or not. The corporate, IoT device, and payment networks must always be kept apart. Owners of smart homes should make sure that every network endpoint is safe and secure. Businesses should periodically conduct in-depth ransomware scans under the guidance of a professional cybersecurity team.

9.3.4   Listening In

IoT gadgets make a great option for listening exercises. IoT devices can be modified by hackers to record audio and video. It is a persistent threat to business, politics, and even the personal life of an individual. The GDP of a nation can be destroyed through industrial and political espionage. Purchase IoT products only from manufacturers who are reputable, branded, certified, and compliant. Continuous monitoring of resources like bandwidth is necessary. Before installation, devices should be extensively examined and inspected for any bugs or electronic surveillance devices.

9.3.5 An Attempt at Social Engineering

Your pet's name is 10 times more likely to elicit a response in an email than a typical marketing email. Hackers spend a lot of time creating a person's profile. They look through the target's social media profiles, business information, blogs (if there are any), contact details, and everything else they may find. They then try to carry out the attack by using this knowledge to convince you that they actually know you. Hackers are extremely cunning and patient. Always request contact information and identification evidence from anyone who sends you an email and claims to know you in person. One should only talk to someone one can remember. Emails used for work, personal, and banking should be separated. It is best to keep personal and professional laptops separate.

It is important to be aware of the potential security risks associated with the use of IoT devices in surveillance and security. By taking steps to mitigate these risks, organizations can help to protect their data and their people. Some of the best practices for securing IoT devices include:

- *Use strong passwords and security settings:* IoT devices should be configured with strong passwords and security settings. This will help to protect them from unauthorized access.

- *Keep devices up to date:* IoT devices should be kept up to date with the latest firmware and security patches. This will help to protect them from known vulnerabilities.

- *Use a firewall:* A firewall can help to protect IoT devices from unauthorized access.

- *Use a VPN:* A VPN can help to protect IoT devices from data breaches.

- *Use a security solution:* A security solution can help to detect and prevent malware attacks.

By taking these steps, you can help to protect your IoT devices and your organization from security breaches.

The majority of IoT security problems may be prevented with proper instruction, knowledge, and increased awareness. However, this does not preclude hackers from identifying weaknesses. The cybersecurity team should be ready with efficient tactics to combat any form of cybersecurity threat, whether it be a DDoS attack or a zero-day exploit.

9.4　IoT SECURITY GUIDELINES

Only certified IoT devices should be purchased, whether one is buying off-the-shelf from a neighborhood electronics store or placing an order on Amazon. The recommendation of a set of protocols to make an IoT device secure before it enters the market has been made by various international agencies and trade groups (both governmental and nongovernmental).

Manufacturers and developers can adhere to cybersecurity standards in accordance with the selected framework depending on their region and the applicable local regulations.

Following are the practically available IoT security frameworks.

A collection of 65 security protocols and practices are advised in ETSI EN 303 645 [1], which is geared toward businesses engaged in the production and development of consumer-based IoT devices.

IoT Security Framework by the National Institute of Standards and Technology (NIST): The NISTIR 8259 Series (for manufacturers), SP 800-213 Series (for federal agencies), and EO 14028 (for consumer IoT devices) are series of standards that the NIST developed to help manufacturers and developers of IoT devices work toward establishing higher security standards for the devices.

U.S. Federal Trade Commission (FTC): The U.S. FTC seeks to enact laws that developers and manufacturers must obediently follow. The security posture of IoT devices across the country is enhanced by this.

In order to support IoT security compliance standards in Europe, the European Union Agency for Network and Information Security

(ENISA) intends to establish the groundwork for next initiatives and advances in the cyberspace. More information about these frameworks is available on the official government websites.

There are 11 crucial conditions for a secure IoT device ecosystem:

- Complex and distinctive passwords;
- No standard or default passwords;
- Security patching and ongoing software updating;
- Encrypted channels for communication;
- Reducing the assault area;
- Dependable backup and resilience strategy;
- 100% device management and visibility;
- Separate corporate and IoT networks;
- Implementing a two-factor authentication and zero-trust policy;
- Microsegmentation implementation;
- Recurring system-wide vulnerability assessments and checks.

The sobering truth that cybersecurity risks are real and that their company might be the next target is dawning on millions of businesses. Recent years have seen significant budgetary investments in cybersecurity and IoT security. This has provided access to a sizable market that is expanding quickly. The IoT security market is anticipated to develop at a compound annual growth rate (CAGR) of 21.2% from 2022 to 2029, reaching $59.16 billion by that year.

9.5 ENTERPRISE IoT SECURITY IN NORTH AMERICA: CURRENT SITUATION

Over 50% of the global market for IoT security is made up of companies based in North America. There is a greater cybersecurity risk as a result of the thousands of North American businesses providing digital services to their international customers. In 2020, more than two-thirds of U.S. organizations paid a ransom to defeat a ransomware assault, according to a Statista IoT security industry report [2]. Companies spent $8.4 million on average in 2020 after a security breach was reported, a 235% or more increase from 2006, when the average cost was $3.54 million. The average cost to organizations worldwide following a security breach was $3.86 million. These IoT security statistics can be difficult to understand, but they accurately depict the situation of business in North America, and the United States in particular.

It is hardly surprising that more IoT devices than PCs are present on the networks of more than 65% of businesses. Additionally, due to the idea of the labor division, the IoT skill gap is particularly noticeable in large businesses. Hacking groups target businesses in the service sector, the healthcare industry, grocery stores, payment gateways, and other sectors where applications are powered by massive amounts of user data. The primary system is accessible to third-party heating, ventilation, and air conditioning (HVAC) system service companies and others. As it is less expensive to maintain than it is to train or recruit skilled employees, this is done to reduce operational costs. Hackers may get access to the primary network through a security flaw on the part of a third-party service provider.

Though businesses claim they are confident in their IoT security implementations, most of them don't have complete control over their IoT devices' visibility and management. If such establishments are penetrated, it would take weeks or even months to locate the hacker's point of access. Only a select number IoT device developers and manufacturers adhere to the numerous standards and rules accessible to them. This noncompliance frequently lowers the product's price on the open market, enticing many businesses. Governments should enact strict rules to ensure that IoT development adheres to particular industry standards in order to put an end to this.

9.6   VIEWING IoT SURVEILLANCE'S FUTURE

The potential uses of IoT surveillance systems are limitless in the future. One of the most exciting IoT advancements coming up is computer vision, in particular. AI is used in computer vision to perform the same type of visual analysis on an image that a human may typically perform. The idea of AI is so well-known that minimal explanation is needed. AI is being used for many jobs that formerly required human intelligence and intervention. For instance, when CCTV first began, a person had to review the video. Many surveillance systems now automatically perform unmanned analysis. Virtually everything may be captured, analyzed, identified, and categorized by IoT cameras that can also warn other IoT system components as necessary. For instance, smoke can be seen by a camera with embedded computer vision software before it reaches a smoke detector and sets off a fire alarm. Or, to assist limit access to a certain location, a surveillance camera can be programmed to detect the faces of specific people.

9.7   FUTURE OBSTACLES FOR IoT VIDEO SURVEILLANCE

IoT video surveillance undoubtedly has a bright future, but there are also some difficulties. The time lag between an issue being detected by a surveillance camera and other systems or personnel receiving an alarm is known as latency, and it is the one thing holding back IoT surveillance solutions. This is due to the fact that many surveillance systems employ a client-server architecture that requires the video stream to pass through at least one cloud server before it can reach other components of your system. However, latency is greatly reduced when remote accessing a security camera via peer-to-peer (P2P) connectivity. With P2P, there is no need to go via a server in order to link the two peers (the camera and the end-user device, such as a smartphone). As a result, the data stream moves through the security cameras much more quickly, improving user experience.

The IoT offers intriguing new possibilities for video surveillance. It enables security providers and manufacturers to stay at the forefront of cutting-edge surveillance technology and offer improved levels of security whenever and wherever needed. The expense and work required to install this crucial technology are more than offset by the advantages. Our lives are becoming more and more entangled with digital devices and environments with each passing second. Our digital interactions will only become more complex as a result of the upcoming metaverse revolution. We are continuously at risk of cyberattacks due to the nonstandard manufacturing of IoT devices and the vast amounts of data they transmit. The demand for IoT security solutions is increased by hazards associated with the use of IoT devices, such as vulnerabilities, cyberattacks, data theft, and other risks.

9.8   WHY ARE IoT SECURITY SOLUTIONS NECESSARY FOR TODAY'S NETWORKS?

When discussing IoT security solutions, there is a compelling case to be made for the lack of physical boundaries, inadequately designed systems, nonstandard device manufacturers, and poor quality control (QC) and quality assurance (QA). Instances of the need for IoT security solutions are listed in Sections 9.8.1.1 to 9.8.1.3.

9.8.1   Securing a Network's Operation and Digital Border

In the context of IoT devices, securing a network's operation refers to protecting the underlying infrastructure that enables communication and data transfer between devices. Traditional network security measures may not be sufficient for IoT environments due to the unique challenges posed by the large number of interconnected devices and their diverse nature.

One significant challenge is the lack of physical boundaries in IoT systems. Unlike traditional networks that are often contained within a physical location (e.g., a corporate office), IoT networks can span across various locations and environments. This lack of physical boundary makes it more difficult to establish and enforce traditional security measures like firewalls and physical access controls. Additionally, inadequately designed systems in IoT deployments can introduce vulnerabilities. IoT devices often have limited computing resources and may prioritize functionality and connectivity over security. This can lead to poor authentication mechanisms, insecure communication protocols, and inadequate encryption practices, making the devices and the network more susceptible to attacks.

Moreover, the proliferation of nonstandard device manufacturers in the IoT ecosystem can further exacerbate security risks. These manufacturers may not follow standardized security practices or adhere to industry guidelines, resulting in devices that are more susceptible to exploitation. It becomes challenging to maintain a consistent level of security across a diverse range of IoT devices from different manufacturers. Furthermore, poor QC and QA practices can contribute to security vulnerabilities in IoT systems. Inadequate testing, lack of security audits, and insufficient monitoring and patching processes can leave devices and the network exposed to potential threats. It is crucial to establish robust QC and QA procedures to identify and address security weaknesses in IoT deployments.

To address these challenges and secure the network's operation and digital border in IoT environments, IoT security solutions are necessary. The block diagram in Figure 9.1 shows the different security measures that can be implemented to secure IoT environments. The measures are divided into three main categories: network security, device security, and user security. Network security measures are designed to protect the network from unauthorized access and to prevent the lateral movement of threats. These measures include network segmentation, strong authentication and access controls, and secure communication protocols. Device security measures are designed to protect IoT devices from unauthorized access, modification, or destruction. These measures include device management and monitoring, regular updates and patching, security audits and assessments, and education and awareness. User security measures are designed to protect users from phishing attacks, malware infections, and other

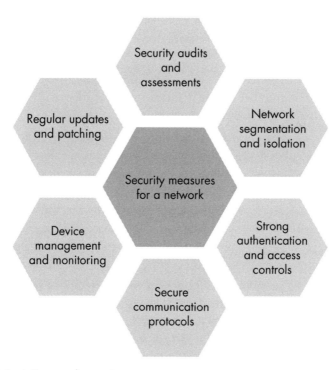

Figure 9.1 IoT network security measures.

threats. These measures include strong passwords, MFA, and security awareness training. By implementing these security measures, organizations can help protect their IoT environments from unauthorized access, modification, or destruction. The following includes a more detailed explanation of each of the security measures shown in Figure 9.1. These solutions typically involve a combination of measures such as:

- *Network segmentation and isolation:* Dividing the network into separate segments to contain potential attacks and limit the lateral movement of threats.

- *Strong authentication and access controls:* Implementing robust authentication mechanisms to ensure that only authorized devices and users can access the network and its resources.

- *Secure communication protocols:* Using encrypted and secure communication protocols to protect data transmitted between IoT devices and the network.

- *Device management and monitoring:* Employing centralized management systems to oversee the security of IoT devices, monitor their behavior, and detect anomalies or potential security breaches.

- *Regular updates and patching:* Establishing processes to regularly update and patch IoT devices to address known vulnerabilities and protect against emerging threats.

- *Security audits and assessments:* Conducting periodic security audits and assessments to identify vulnerabilities, assess risks, and implement necessary security improvements. Promoting security awareness among IoT device manufacturers, network administrators, and end users to foster a security-conscious mindset and best practices.

9.8.1.1 Data Security

Data security is another critical aspect of IoT security solutions. IoT devices generate and transmit vast amounts of data, often including sensitive information. This data can include personal data, operational data, or even confidential business information, depending on the specific IoT deployment.

Ensuring data security in IoT environments involves safeguarding the confidentiality, integrity, and availability of the data. The diagram in Figure 9.2 shows that data security in IoT environments involves safeguarding the confidentiality, integrity, and availability of the data.

This can be achieved by implementing a variety of security measures, including encryption, access controls, data privacy, data integrity and tamper detection, data storage and transmission security, data lifecycle management, data access monitoring and auditing, user authentication and authorization, secure data sharing, secure data backups and disaster recovery, security incident response, continuous monitoring and threat intelligence, and employee training and awareness.

Some key considerations are as follows:

- *Encryption:* Implementing strong encryption techniques to protect data both at rest and in transit. This ensures that even if the data is intercepted, it remains unreadable to unauthorized individuals.

- *Access controls:* Implementing access controls and role-based permissions to ensure that only authorized individuals or systems can access and manipulate the IoT data.

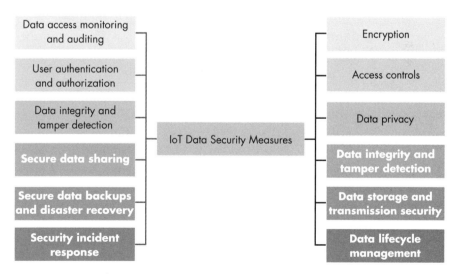

Figure 9.2   IoT data security measures.

- *Data privacy:* Complying with relevant data privacy regulations and guidelines, such as the General Data Protection Regulation (GDPR), to protect and individual's privacy rights and handle personal data appropriately.

- *Data integrity and tamper detection:* Implementing mechanisms to ensure the integrity of IoT data, such as digital signatures or checksums, to detect any unauthorized modifications or tampering attempts. This helps ensure that the data remains intact and trustworthy throughout its lifecycle.

- *Data storage and transmission security:* Implementing secure protocols and practices for storing and transmitting data. This may involve using secure cloud storage solutions, encrypted communication channels, and secure application programming interfaces (APIs) protect data while it is in transit or at rest.

- *Data lifecycle management:* Establishing proper procedures for managing data throughout its lifecycle, including data collection, storage, usage, sharing, and disposal. This involves defining data retention policies, securely deleting data when no longer needed, and ensuring proper data anonymization or pseudonymization techniques are used when necessary.

- *Data access monitoring and auditing:* Implementing mechanisms to monitor and log data access activities, including who accessed the data, when, and for what purpose. Regular auditing of data access logs helps detect any suspicious or unauthorized activities and enables quick response and investigation.

- *User authentication and authorization:* Implementing strong authentication mechanisms, such as two-factor authentication or biometric authentication, to verify the identity of users accessing IoT data. Additionally, enforcing granular authorization controls ensures that users can only access the data they are authorized to view or manipulate.

- *Secure data sharing:* If data needs to be shared with external parties or across different systems, implementing secure data sharing practices is essential. This may involve using secure data exchange protocols, establishing secure APIs or data interfaces, and implementing proper access controls and encryption for shared data.

- *Secure data backups and disaster recovery:* Implementing regular data backups and establishing a robust disaster recovery plan to ensure data can be restored in case of system failures, data breaches, or natural disasters. Backup data should be securely stored and protected to maintain its confidentiality and integrity.

- *Security incident response:* Establishing a well-defined security incident response plan to address data breaches or security incidents promptly and effectively. This includes defining roles and responsibilities, establishing communication channels, and conducting regular incident response drills to ensure a timely and coordinated response to security events.

- *Continuous monitoring and threat intelligence:* Implementing continuous monitoring of IoT devices and the network infrastructure to detect any abnormal behavior or security threats. This can involve using intrusion detection systems (IDS), security information and event management (SIEM) tools, and threat intelligence feeds to stay informed about the latest security threats and vulnerabilities.

- *Employee training and awareness:* Conducting regular training and awareness programs for employees to educate them about data security best practices, the risks associated with IoT devices, and how to handle data securely. This helps create a security-conscious culture and reduces the likelihood of human error leading to data breaches.

By considering these aspects of data security and implementing appropriate measures, organizations can enhance the protection of IoT data, mitigate the risk of data breaches, and maintain the privacy and integrity of sensitive information.

9.8.1.2   Typical Threats to IoT Devices

The variety of threat vectors that IoT devices frequently encounter is their biggest problem. A few can be attributed to manufacturers and firmware developers, but others may be the result of system exploitation and focused cyberattacks. The growing number of IoT devices in homes has led to increased concerns about cyber intrusion. Many homes lack IoT security solutions to protect their data.

Using outdated operating systems, hackers can access networks. IoT devices with obsolete or unsupported operating systems are easily vulnerable. By breaking into only one weak system on the network, hackers can bring the entire system to a halt. 300,000 Windows-based computers were targeted by the 2017 WannaCry ransomware. Systems with outdated security were successfully breached.

Inadequate testing and encryption. Inadequate testing and encryption are the results of poor QC and QA. The network is susceptible to attacks when there are no IoT security solutions added to it. Eavesdropping has become a profession as high technology has been more widely available.

Service ports that are open (Teletype Network (Telnet) and Secure Socket Shell (SSH)). According to a story on ZDNet in 2020 [3], when the Telnet ports were left open, a hacker leaked the credentials of over 500,000 IoT devices, residential routers, and servers. Similar to this, the National Exposure Index published by Rapid7 in 2017 [4] asserted that over 10 million IoT and other devices have their Telnet ports exposed. Following the deployment of the product, the development teams should close the Telnet ports.

Access HVAC and other systems. The largest threat to IoT networks is access through HVAC and other remotely operated systems. Vendors frequently receive remote access to install systems and firmware. Strong firewalls and IoT security tools frequently do not protect the endpoints of vendor systems. This is viewed by hackers as a gateway to the entire IoT network.

9.8.1.3 *IoT Networks That Are Most Susceptible to Hackers*

Each IoT network has a variety of IoT security tools set up at different levels and points of failure. Frequently, the most impacted IoT networks are the commercial, consumer, and medical ones. There are too many failure spots in a consumer IoT network. The most susceptible spots are those with outdated operating systems and default passwords. Remote access suppliers of unmanaged IoT devices are frequently the root cause in commercial IoT networks. Threats to consumer IoT and commercial IoT networks frequently arise due to affordability (in the case of consumer IoT devices) and inadequate security testing. The issues the minerals and mining business faces include equipment from different vendors running different operating systems and unsupported/outdated operating systems.

Although businesses and individuals have implemented a number of IoT security solutions, hackers are still able to access networks through IoT devices and trigger cascade consequences. These networks are frequently the easiest to hack if they lack reliable security measures and real-time management. Insiders provide solid evidence in favor of IoT security solutions! Many industries are concerned about employee spying. There are reports that insiders intend to introduce ransomware into networks, giving hackers unauthorized access to sensitive data. In 2019, Tesla's systems were almost targeted by a corrupted malware attack, but the employee involved changed their mind. Enterprises must improve how they restrict access to sensitive information while maintaining knowledge transfer and other aspects of production. This reveals a completely new facet, namely the requirement to safeguard data even when internal systems are vulnerable. IoT security solutions come into play in this scenario and frequently save numerous businesses.

9.9 HOW SHOULD WE HANDLE DATA PRIVACY?

Every node of the digital world is in danger due to the easier access to cutting-edge technology. When there is a data breach, businesses pay dearly because regulators begin to doubt their security system. As a result, choosing security is determined by one's need for privacy. While it is virtually impossible to keep track of every person involved in an IoT device's manufacturing process, picking the best provider of IoT security solutions shouldn't be too tough. It is too late to be discovered.

According to IBM's "Cost of Data Breach" report [5], it takes an average of 287 days, or more than 9 months, before a cyberattack is discovered. By the time the intrusion is discovered, hackers frequently get access to crucial data, escalate their privileges, and even export sensitive information.

It is a shock to learn that only 2% of the traffic passing via IoT devices is encrypted, with the remaining 98% being unencrypted [6]. Businesses cannot operate while entrusting anything to unknowable individuals. It is insufficient to only protect IoT devices. The best course of action in the current situation is to protect critical data even when the systems are compromised. Not all detections result in protection. The majority of security systems generate 5,000 to 10,000 warnings every day. There is far too much information here for someone to monitor and go through each warning. Because of this, hackers' jobs are made simpler. The production manager for a security company, Eyal Arazi, asserted in an interview with ZDNet that "current security systems definitely detect too much" and informs you of the actual situation on the ground. The major problem with most IoT security solutions is that the alarms they produce lack context and correlation. Enterprises should ask the provider if the security suite has the ability to automatically correlate individual events and create a logical sequence out of them.

9.10 CONCLUSION

In conclusion, vulnerabilities and threats continue to evolve, and so should IoT security solutions. IoT device threats are real and require prompt attention. Businesses should be aware of the security precautions taken by their suppliers. Some of the security measures that businesses can take to protect their IoT devices include establishing separate networks for corporate and IoT device management, restricting access to third-party vendors, requiring cryptographic hash keys for vendors to sign onto the network, scheduling constant monitoring and tracking of IoT devices, and implementing IDS and Intrusion Prevention System (IPS). The hyperadoption of IoT devices is greatly expanding the attack surface, putting security at the mercy of hackers. To guard against, stop, and deter cyber threats, businesses should embrace strong IoT security solutions right away.

In addition to the security measures mentioned, businesses can also take advantage of the security features of antennas to improve the security of their IoT devices. For example, antennas can be used to encrypt data transmissions, filter out unwanted signals, and detect and block unauthorized access. By using antennas to improve the security of their IoT devices, businesses can help to protect their data and assets from cyber threats.

References

[1] ETSI EN 303 645:2020, Cyber Security for Consumer Internet of Things (IoT): Baseline Requirements.

[2] "50 Ransomware Statistics and Latest Trends," Statista, February 14, 2022.

[3] Leroux, J., "Why You Should Never Use Telnet and How to Secure Your SSH Server" ZDNet, March 10, 2020.

[4] "Why You Should Stop Using Telnet" by Rapid7, February 20, 2017.

[5] "Cost of a Data Breach Report," IBM: IBM Security, July 24, 2023.

[6] "Expanding IoT Visibility," Palo Alto Networks, October 19, 2020.

10

RELIABILITY AND SECURITY CHALLENGES

10.1 INTRODUCTION

Reliability is a crucial but previously underappreciated feature of IoT systems. This is due to the fact that IoT systems are being used in critical applications, including those in which the safety of people is at risk. The difficulties and particular requirements will be covered in Section 10.2. Antennas are essential components of IoT systems, as they are responsible for transmitting and receiving wireless signals. However, antennas can also be a source of reliability and security challenges. Because they are no longer considered toys and are used in applications with moderate to high criticality requirements, IoT systems require reliable architecture, operation, and application development.

10.2 UNAVOIDABLE ENCOUNTERS

The concept of reliability requires that errors in the software, hardware, interactions with the real world, and user interactions all be taken into consideration. One of the changes is that the systems now regularly produce large quantities of data at high rates, which places additional strain on the mechanism responsible for reliability. The data may be of mixed criticality, which means that some of it may be critical and, if improperly handled, result in user-visible failures, while the rest of it may not. This makes heterogeneous reliability processing necessary to handle the data properly. The heterogeneity of our target systems is the first-order characteristic which is applicable to the reliability domain in terms of both design and operation. For instance, the software on certain devices is developed using stringent software development procedures and programming languages that are safe by design. On other devices, the software may be developed using agile software development techniques and programming languages that are unsafe. Additionally, due to the fact that runtime instantiation occurs differently on each device, the error tolerance levels can vary greatly. For instance, there may be some components that are able to hide problems while others are responsible for spreading them. In conclusion, dependability measures are unable to considerably affect the timing of the procedure because it is a real-time process. While the design and development of reliability in hard real-time systems is well-established, our target systems present new challenges because they are produced more quickly (for example, with very little to no formal validation) and operate in environments that are more diverse and uncertain than hard real-time systems. The concept of predictable behavior from the system, despite the presence of several unpredictable circumstances, is related to the subject of reliability. These factors include interactions within the IoT platform, interactions throughout the platforms, and interactions between the system and human users. Because IoT systems commonly incorporate human-in-the-loop (HITL) or human-on-the-loop components, this is extremely important (the former means a human has to be involved in the chain of decision making while the latter makes that optional). Different individuals have varying levels of tolerance for ambiguity, which underscores the need of including this component in the functioning of an IoT system.

One of the biggest challenges with antennas in IoT is that they are susceptible to environmental factors such as weather, interference, and noise. These factors can cause signal attenuation, which can lead to data loss or communication errors. Antennas can be affected by a variety of weather conditions, including rain, snow, fog, and wind. These conditions can cause signal attenuation by absorbing or scattering the radio waves. They can also be affected by interference from other sources, such as other wireless devices, power lines, and lightning. This interference can cause signal attenuation or distortion, which can lead to data loss or communication errors. Antennas can also be affected by noise, which is any unwanted signal that can interfere with the desired signal. Noise can come from a variety of sources, including electrical devices, lightning, and the atmosphere.

These environmental factors can cause signal attenuation, which is a decrease in the strength of the radio signal. Signal attenuation can lead to data loss or communication errors.

There are a number of ways to mitigate the effects of environmental factors on antennas. One way is to use a directional antenna that can focus the radio waves in a specific direction. This can help to reduce interference from other sources. Another way to mitigate the effects of environmental factors is to use a high-quality antenna that is designed for the specific environment in which it will be used.

10.3   STANDINGS AND SITUATIONS

It is necessary for there to be a wide variety of dependability protocols since this reflects the wide variety of runtimes on which they will operate as well as the applications that they are designed to protect. The reliability protocols should be able to adapt to the device's current resource status (e.g., a resource-intensive but crucial task may begin running on it), the device's current reliability requirement (e.g., the data stream that is currently being gathered, processed, and communicated to the back end may be highly critical for some downstream application), and the device's current physical environment (e.g., a physically hazardous environment may cause correlated failures of multiple devices in spatial proximity).

For antennas, the challenge is that they can be difficult to position in a way that maximizes signal strength and minimizes interference. This is especially true for IoT devices that are located in remote or challenging environments.

10.4 RESEARCH CONCERNS

Researchers are working on developing new antenna technologies to overcome the challenges that affect the reliability of IoT systems. There are four primary domains, as seen in Figure 10.1, that can be used to classify the primary research problems that have an effect on the reliability of IoT systems.

1. *Management of correlated failures*: It is necessary to deal with failures that are linked to one another in place and time in order to go forward. The phenomenon known as spatial correlation arises when many devices have a chance to experience similar actual or virtual conditions at the same time, such as congestion in wireless networks or swings in temperature. When a physical phenomenon spreads over time and affects devices in a sequential manner, this is an example of temporal correlation. Other examples of temporal correlation include when a device fails as a result of high moisture con-

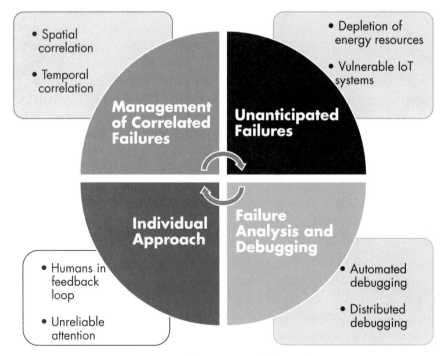

Figure 10.1 Research concerns affecting the reliability of IoT systems.

tent or when a large group of people move in unison, producing an excessive number of concurrent occurrences.

2. *Handling failures that were not anticipated*: Unpredictability accounts for a sizable fraction of system failures in any system. This effect is magnified in IoT systems for a variety of reasons. To begin, the energy sources are unexpectedly depleted. This may be the result of external variables in the case of rechargeable solar batteries or an unexpected load that generates a high level of communication activity. Second, the deployment of an IoT system does not have a lot of built-in safety features, which means that it does not have a lot of headroom. As a result, even conditions that are only somewhat out of the ordinary, such as momentary spikes in load, have the potential to throw the system into disarray and cause it to collapse. Third, the failure modes of these systems are not modeled as accurately as the failure modes of server-class systems are.

3. *Failure analysis and debugging*: It is essential to provide automatic failure analysis and debugging in IoT systems. This is due to the fact that the systems are composed of a large number of disparate devices that would place a significant burden on human intellect if they were to be debugged. Automated debugging is challenging to accomplish because not all execution data can be captured at the devices, and not all logged data can be communicated to the back end for debugging. In addition, when performing distributed debugging, it is frequently essential to integrate the traces from a number of different devices.

4. *Taking into account individuals*: This problem with dependability is caused by the fact that many of these IoT devices incorporate humans in the feedback loop (or on the loop). This has a variety of repercussions for a wide range of applications and even for different configurations of the same program. For example, certain human users may have a high aversion to the rate of false alarms, while others may find it revolting to observe alarms on small-form factor displays that are found on devices. A typical form of unreliability that is human-centric emerges whenever human users are busy do-

ing other things while interacting with technology. The question of maintainability is inextricably tied to this overarching subject in order for these systems to be maintained (updated, reflashed, reconfigured, etc.) with little to no human intervention and scarcely any intervention from an expert.

Researchers are working on developing new antenna technologies to overcome these challenges. For example, researchers are developing antennas that are more resistant to environmental factors and that can be more easily positioned in challenging environments. They are also developing antennas that are more efficient, which can help to extend the battery life of IoT devices. The development of new antenna technologies is an important part of the effort to make IoT systems more reliable. By overcoming the challenges that affect the reliability of IoT systems, researchers can help to make IoT a more powerful and widespread technology.

10.5   STABLE FOUNDATIONS UPON WHICH TO CONSTRUCT

For each of the aforementioned topics, there is very little to some ongoing research and development. In this section, we take a look at some of the more interesting aspects of each research area. In order to make IoT systems more reliable, experts have first created a plethora of methods to locate faulty sensors and architecture. The investigation that is described in [1] makes use of the realization that fundamental detectors would detect a number of events that are practically contemporaneous with one another in time when there are related failures. The authors demonstrate that a single-level ML classifier underperforms for many practical system-level problems, but two-stage detection (clustering events in the first stage) greatly reduces false-positive rates and significantly enhances detection rates. Bychkovskiy et al. [2] proposed a technique for large-scale sensors that involves two phases of postdeployment calibration in order to overcome the issue of space-correlated failures. The fundamental idea here is to achieve the highest possible level of consistency across many groups of sensor nodes by capitalizing on the temporal correlation that exists between the data collected by colocated sensors. Blind calibration for sensor networks using values that are poorly correlated across sensors was a concept developed by Balzano and Nowak [3]. On the other hand, Neumayer and Modiano [4] developed tools to model and analyze regionally associated network failures while focusing on network connection. These tools were established to model and analyze regionally associated network failures.

In the event that temporally related failures occur, Sharma et al. [5] advised using approaches that are based on time-series analysis to locate damaged sensors. Extensible sensor stream processing (ESP) is a framework that was introduced by Jeffery et al. [6] to clean time- and space-correlated sensor data that is captured on devices when the system is operational, and the traces are somehow returned to a backend for replay and debugging. Embedded system profiler was designed to clean time- and space-correlated sensor data captured on devices in real time. However, it is only applicable to a single node, and it does not take into account the execution on microcontrollers that are frequently utilized (e.g., those which run multithreaded operating systems (OSes) and applications). The research carried out by Aveksha [7] makes use of supplemental hardware in order to record Joint Test Action Group (JTAG) port traces without interfering with the operation of the node; nonetheless, it is unable to record complete control flows. Both this technology and other methods that utilize hardware modifications are incompatible with commercial off-the-shelf (COTS) IoT systems and hence cannot be deployed. Some software-only initiatives, such as TinyTracer and Prius, which selectively record some events, do not provide replay-based debugging and so cannot be used (just control flow for TinyTracer). The questions that need to be answered include how to provide high fidelity system-level replay, which is defined as a replay that can accurately replicate control flow at the instruction level and the state of memory at any given instant for any software module that is running on the node. This is a challenge that has not yet been answered. The problem of detecting patterns in the traces, which is broader, is related to the fact that there must be learning algorithms that are capable of learning these patterns from field observations. These methods will take the place of the current rule-based ones, which are fragile. Concerning the topic of human aspects in dependability, researchers have developed strategies to improve the dependability of systems and identify faults as soon as they occur.

A tandem human-machine cognition strategy was used by Gross et al. [8] to reduce and avoid cognitive overload situations where false alarms and ambiguity may overwhelm humans as human operators get involved in the control loop of IoT sensor networks. This cognitive overload situation can occur when humans become involved in the control loop of IoT sensor networks. The way in which humans interact can also have an effect on the interaction that exists among sentient entities. Guo et al. [9] developed opportunistic IoT models in order to facilitate information sharing and dissemination across opportunistic IoT communities that are generated based on human activity. This was done in order to facilitate information sharing and dissemination across opportunistic IoT communities. Szewczyk et al. [10] learned from their experience with a sensor network expedition that the failure of temperature sensors is closely connected with the failure of humidity sensors in addition to failures that are related to space and time. This discovery was made while the researchers were analyzing their findings. This data from linked sensors stands to gain a great deal from the analytic models being developed by experts in the field of data mining. In the context of truth discovery, where contrasting information can come from a range of sources, Dong et al. [11] took into account the degree to which different data sources depended on one another. There are a number of challenges that need to be addressed before ML-based, correlation-based failure detection can be widely adopted. One challenge is the need for large amounts of labeled data to train ML models.

Another challenge is the need to develop ML models that are robust to noise and other sources of variation in sensor data. As a continuation of the topic of dealing with unforeseen failures, a series of solutions have been applied to energy-harvesting IoT devices. In these types of devices, failure can occur unexpectedly due to energy drain. In a number of the works that have been published in this area, checkpoints allow the application to save and restore its prior state. Some sophisticated work does the check pointing based on the amount of energy that is available. A few researches that were done recently give lightweight solutions. Karimi and Kim [12] presented a new energy scheduling strategy to carry out periodic real-time operations on the embedded systems that are only powered by an intermittent source of energy. Additionally, Maeng and Lucia [13] presented the adaptive low-overhead scheduling for intermittent execution. The unsolved questions, on the other hand, are how to manage an increasing number of unanticipated setbacks while at the same time showing respect for the resources that are already available (available storage, energy, etc.). Furthermore, the most compelling research on the subject of fixing bugs focuses on the collection of runtime data and the automatic determination of abnormal behavior by mining data patterns. Sadly, there aren't many practical ways that may be used to debug issues that occur during production. Record and replay is one potential route in which execution traces are stored on a service platform with a trust model that incorporates both reputation-based properties and knowledge-based properties, so that different entities can trust one another. This model enables the recording and replaying of execution traces. On the other hand, Cranor [14] presented a paradigm for thinking about the HITL within the secure system in order to find potential causes for human failures. However, there are still questions that haven't been solved on how to use a complete model to investigate how different demographics of people influence the dependability of IoT systems, possibly with ML models.

It is vital to have an understanding of the definition of IoT in order to have an understanding of the dependability problem that exists within the paradigm. IoT definitions are frequently hazy and underrepresented. It is common practice to roughly define the IoT as the capacity to connect commonplace objects to the internet, thereby making it possible for your toaster to "talk" to your refrigerator. This statement may be true for a portion of the IoT paradigm, but it does not cover the entire IoT paradigm. One can begin to define the IoT paradigm by first considering the fundamental components of the IoT. Among these are detecting, acting on, communicating, offering services, and applying. The architecture of the IoT depicted in Figure 10.2 can then be applied to these four components.

The sensing and acting operations are carried out on the device layer, which is the lowest layer of the architecture. The following layer above, known as the edge layer, is responsible for making it feasible for devices and the application layer to communicate with one another. In most cases, semicapable devices functioning as hubs make this connection possible by collecting data from the sensors, transmitting it into the cloud, and giving commands to the actuators as necessary. Keeping these essential aspects in mind, the IoT can now be defined in its entirety as a paradigm that enables interconnectivity in everything to establish monitoring and control infrastructure that can

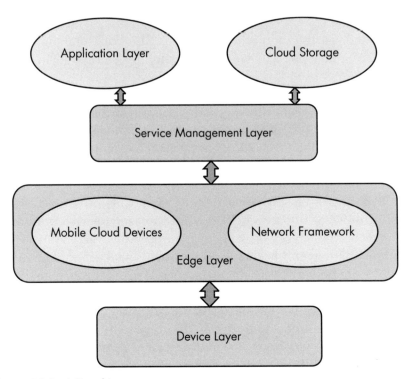

Figure 10.2 IoT architecture.

be used in applications to improve user experience on a day-to-day basis.

10.6   DEVICE DEPENDABILITY

The extremely restricted functionality of these devices, which include the sensors and actuators, is likely to be the first problem that we come against. These restrictions pertain to compute, memory, and the available capacity of the battery. IoT apps have an issue with batteries because it can be difficult to determine when a device needs a new battery since the application layer is frequently unaware of how much battery is still in it. This makes it difficult to determine when a device needs a new battery. The issue of the length of time a battery can last becomes even more of a concern when we consider the fact that some electronic devices may be put in difficult or dangerous environments. Because of memory and processing power constraints, IoT devices can only store simple encryption methods, so the only way to protect the data that is being conveyed by the device is to utilize lightweight encryption. Another issue arises as a result of the restricted design of the devices when it comes to the process of updating the limited firmware of these sensors that only have a low amount of power. Due to the lack of power and the potential impact on the device's battery life, connecting to a cloud service on a frequent basis to assess whether new firmware has to be downloaded and installed on the device is unfeasible.

Because of this, there is a possibility that some devices will be running outdated firmware that leaves them open to various forms of cyberattacks. The sensors and actuators that are utilized by the IoT are typically located in remote locations, where they run the risk of being subjected to unfavorable climatic conditions such as extremes of heat and cold, mechanical wear, vibration, and dampness. The useful life of a device needs to be calculated in order to ascertain the appropriate time for the item to be deactivated. Because of this, we could expect observing considerable differences in device lifetime for identical devices used in different settings, which makes it difficult to monitor the system's reliability. The device's useable life will be shortened if it is used in a harsh environment. Another concerning aspect of the dependability of IoT devices is the propensity of sensors to *fail-dirty*. The phrase *sensor failure with continued false signal transmission* refers to the phenomena that occurs when a sensor fails yet continues to send out erroneous readings. This is a widespread problem in the configurations of IoT devices that is well-known but poorly understood. In this particular case, the sensor appears to be operating normally, which makes it challenging to determine the issue. When we consider the fact that actuation frequently has a direct and tangible impact on the lives of people, it becomes clear that the consequences of sending a false reading in the context of the IoT could be severe.

10.7 THE DEPENDABILITY OF NETWORKS AND COMMUNICATION SYSTEMS

Mobility is one of the primary needs for an IoT network that enables users to transition between applications while the device onboarding and identification process proceeds smoothly in the background. This is one of the fundamental requirements. However, because there is a lack of cooperation among manufacturers in the provision of globally unique identities for all IoT devices, global addressing in the IoT presents a challenge. This suggests that the IoT network itself is in charge of doling out distinctive identities to devices. Because it is expected that IoT devices will be mobile, this presents a challenge since the identifier of the device can change between different networks. If this happens, it is possible that device traceability would be lost. This creates a dependability problem when trying to monitor or audit the device as it moves across different IoT applications. Internet Protocol is the current standard that is used as the de facto benchmark for identification and communication in conventional networks. However, Internet Protocol in its current form is not a particularly good fit for the IoT. When new protocols are introduced into this problem area, which is not always an easy task, rapid maturation will be required of them as quickly as possible. When implications of using unique addressing are taken into account, the scenario becomes considerably more precarious. When considering the estimations of 50 billion devices and the 32-bit length of IPv4's addresses, which allows for 4.3 billion addresses, it is obvious that Ipv4 is not sufficient to realize the potential of the IoT. This problem is made considerably more difficult by the fact that Ipv4 exhausted its available addresses in the year 2010. It is important to create a protocol that has appropriate addressing space. An example of this would be Ipv6, which has a 128-bit address space with room for 3.4×10^{38} different addresses. However, constrained devices have challenges when attempting to use this new addressing space because not all of them are able to manage the overheads necessary for the address. The 6LoWPAN protocol provides a solution to the considerable address overhead. 6LoWPAN has the potential to lessen the amount of space taken up by the header of Ipv6 packets, rendering them more suitable for use with the IoT. This patchwork of inconsistent standards and deployments among IoT devices and deployments meant to facilitate communication for limited devices in IoT networks has resulted from the brand-new and developing standards that were developed to address the IoT's evolving requirements. It is more difficult to evaluate the dependability of the network connection as a result of the fact that not all of these protocols offer QoS guarantees due to the fact that they are lightweight and restricted in nature. Because of the network's low resources and irregular communication, the network is also prone to drop readings or produce incorrect data. It is worrying that measurements could be lost due to the fundamental properties of IoT networks, especially considering that IoT infrastructure is usually responsible for monitoring mission-critical applications.

10.8 AVAILABILITY AND DEPENDABILITY AT THE APPLICATION LEVEL

The application layer of the IoT paradigm does not have to contend with the same constraints as the architecture's network or device layers. It is quite important to keep in mind that the dependability of the application layer is frequently influenced by the dependability of the lower architectural levels. If abnormal data is sent from the device via the network and into the application layer, the program's reliability will suffer as a result. It is essential for the application layer to have sufficient anomaly detection algorithms in order to keep the program's dependability intact and eliminate errors. This process can be difficult due to the huge variety of restricted devices that are included in IoT networks. These devices communicate many pieces of information in a variety of formats. Even though the application layer is not constrained by the same physical constraints as the device layer, it is still vital to manage the reliability of the programs that are being deployed. Some classifiers are more prone to mistakes than others when classifying human activities in IoT applications.

10.9 FINDING EFFICIENT SOLUTIONS TO ENSURE THE DEPENDABILITY OF IoT SYSTEMS

The IoT is vulnerable due to the difficulties in dependability that exist at all three levels of the architecture. These difficulties frequently lead to abnormal data being generated and transmitted over the network. This concept exemplifies the important requirement for effective, quantitative dependability measurements that will help us to determine whether or not our IoT devices are suitable for the tasks for which they were designed. Because judgements will be made based on information that, in the worst possible scenarios, could put human lives in jeopardy, the subject of anomalous data is of the utmost importance for the IoT goal. Any framework that tries to evaluate and measure dependability in the IoT must therefore be able to identify the presence of irregularities inside the system. After dependability has been measured, there will be a fresh opportunity to involve another HITL, which will result in an increase in the robustness of the system. The HITL paradigm creates opportunities for discovering and addressing dependability concerns in essential IoT infrastructure that can then be addressed and resolved. Employing a human observer provides an additional layer of domain specialist knowledge to the application. This enables the human to draw judgements about the dependability of the system by combining the data from the system with their own prior knowledge of the subject matter as an expert. In addition, human beings are capable of evaluating ground truth, such as the actual temperature, that can be used to validate the results of a computer reading. This concept is very new in the field of study on the IoT, and as of yet, no reliability studies have decided to combine conventional reliability models with an HITL method in order to assist with dependability evaluation. This peculiar concoction could be the first step toward developing a new and efficient treatment. This section discussed the most important concepts and prerequisites for achieving reliability in the IoT. In light of the delicate environment that the IoT resides in, the research community has a responsibility to develop and implement theories and solutions that can assist in the process of measuring and comprehending the dependability of our core IoT infrastructure, the application that is more likely to fail. It is imperative that developers make a concerted effort to evaluate and comprehend the dependability of applications that are housed in IoT infrastructure in order to reduce the likelihood of critical mistakes being introduced into the system.

10.10   THE SIGNIFICANCE OF PROTECTING CONNECTED DEVICES

Because IoT technologies are utilized so extensively, businesses need to pay careful attention to their system security. Any vulnerability might lead to a malfunction in the system or an attack by hackers, both of which would have repercussions for hundreds or perhaps thousands of people. For instance, traffic lights could fail, which would lead to car accidents, and burglars could disable a home security system, which would allow them to break in. Because some IoT devices are utilized for human safety or healthcare, the matter of IoT device security may be of the utmost significance to the lives of individuals.

In order to keep their data safe, IoT systems need to give security a very high priority. The large amounts of sensitive data, including personally identifiable information, that smart devices acquire require compliance with a wide variety of cybersecurity rules, standards, and regulations. This is necessary in order to protect the data. In the event that such information is compromised, legal action and/or sanctions may be taken. In addition to this, it can lead to a loss of trust from customers as well as damage to one's reputation. IoT security is a collection of methods and procedures for defending against a variety of IoT security intrusions on the physical objects, networks, operations, and technology that make up an IoT ecosystem. This collection of methods and procedures is known as the IoT security stack.

The two primary goals of IoT security are to:

1. Make certain that all of the data is gathered, processed, stored, and sent in a safe manner.
2. Recognize and fix IoT component vulnerabilities.

Considering the security concerns about the IoT, Kaspersky [15] documented 639 million breaches for the entire year of 2020, but still there were 1.51 billion breaches of IoT devices from January to June of 2021. When developing IoT systems, it is unacceptable to minimize the relevance of ensuring cybersecurity. Prior to gaining an understanding of how to keep IoT systems safe, it is essential to conduct research into the various cybersecurity problems that may arise. The resource constraints and low processing power of many smart devices make it difficult to maintain IoT security. This is the primary reason for the difficulty. Because of this, they are less able to carry out rigorous, resource-intensive security activities, making them more vulnerable than devices that are not connected to the IoT.

Many of the systems that make up the IoT have security holes due to the following reasons: inadequate access control in IoT systems, insufficient processing power for effective built-in security, a lack of funding for adequately testing and enhancing firmware security, and a lack of frequent patches and upgrades as a result of the technical constraints of IoT devices and limited funding sources.

- If users do not upgrade their devices, it may reduce the ability to patch vulnerabilities, and older devices may become unable to get software updates in the future.
- Inadequate defense against physical assaults: a culprit can approach the gadget to the point where they can implant a chip or use radio waves to corrupt it.

An IoT system is a target for hostile actors who wish to infiltrate its communications, introduce malware, and steal sensitive data. They aim to do this in order to get access to the system. For example, hackers were able to access Ring smart cameras by using credentials that were not strong, had been recycled, or were the default setting. They were even able to make remote contact with victims by utilizing the microphone and speakers that were built into the camera.

10.10.1 Unsecured Correspondence

Because of their limited resources, IoT devices make it difficult to implement the majority of the currently available security measures. Because of this, conventional security measures are not nearly as effective as they should be in protecting the communication of IoT devices.

Insecure communication can expose users to a number of detrimental risks, one of the most serious of which is the possibility of a man-in-the-middle attack (MitM). Hackers can easily carry out MitM attacks if your device does not use secure encryption and authentication methods. These attacks allow hackers to breach an update process and get control of your device. Attackers have the capability of doing everything from installing malware to making functional alterations. Even if your device is not the target of a MitM attack, cybercriminals may nevertheless be able to intercept the data it sends via clear text messages with other devices and systems.

When a gadget is connected to a network, it opens itself up to possible attacks from other connected devices. Consider how quickly an attacker can obtain access to all other unisolated devices in a home network if they are only able to gain access to one of those devices.

10.10.2   Data Breaches Involving IoT Systems

It is possible for hackers to acquire the data that is processed by your IoT system by intercepting messages that are not encrypted. It is also possible that this will contain private information about you, such as your location, your finances, or even your medical history. Even though it is not the only method, one way that attackers can obtain useful information is by taking advantage of communications that are not adequately protected. The cloud is used for the transportation of all data as well as its storage, and services run from the cloud are equally vulnerable to attacks from the outside world. As a consequence of this, it is possible for data to be compromised not just on the devices themselves but also on the cloud environments to which they are tied. The use of third-party services presents yet another opportunity for data to be compromised within your IoT devices. For example, it was found that Ring smart doorbells were unlawfully sending customer data to social media platforms like Facebook and Google [16]. This scenario transpired as a direct result of the Ring mobile app having third-party tracking services activated on its end.

10.10.3   Malware Risks

According to the findings of a recent survey conducted by Zscaler [17], the types of electronic devices that are most likely to be targeted by malware are smartwatches, smart TVs, and set-top boxes. If malicious actors are successful in installing malware in an IoT system, the functioning of the system could be altered, personal information could be collected, and additional attacks could be carried out. In addition, if the manufacturers of some devices do not take adequate efforts to ensure the safety of their software, the devices themselves may already be infected with viruses.

The most well-known piece of malware that targets IoT devices has already been combatted in ingenious ways by multiple companies. There is a lesson that can teach you how to defend your systems proactively against the Mozi IoT botnet, and an FBI agent recently detailed how the agency stopped the Mirai botnet attacks. Both of these resources are available on the Microsoft website [18]. However, cybercriminals are continually developing new vulnerabilities that can be exploited via IoT networks and devices. Researchers [19] believe that the virus known as BotenaGo, which was developed in 2021 using the programming language Golang, has the ability to exploit over 30 distinct flaws that are present in smart devices.

10.10.4 Cyberattacks

In addition to the malware and MitM assaults that were discussed in Section 10.10.3, there are a variety of other incursions that may be possible with IoT systems. The following is a list of the most common types of attacks on IoT devices:

1. *DoS attacks*: Because of their limited computational capacity, IoT devices are particularly vulnerable to DoS attacks. This attack affects a device's ability to respond to genuine requests by overwhelming it with fake traffic.

2. *Denial-of-sleep (DoSL) attacks*: Batteries that do not need to be often charged are typically used to power sensors that are connected to wireless networks. This is done so that the sensors can maintain a constant monitoring of their environments. The battery life of the smartphone can be significantly increased by keeping it in the sleep mode for most of the time. The ability to consciously manage when one sleeps and when one wakes up is contingent on the communication requirements of various protocols, such as MAC. Flaws in the MAC protocol could provide an opportunity for an attacker to carry out a DoSL attack. This type of attack depletes the battery, which renders the sensor useless.

3. *Device spoofing*: This exploit can take advantage of a device that is susceptible to compromise because it improperly integrated digital signatures and encryption. For example, cybercriminals may exploit a vulnerable public key infrastructure (PKI) to spoof a network device in order to disrupt IoT deployments and cause other problems.

4. *Invasion in a physical sense*: Even though the vast majority of attacks are carried out remotely, it is still possible to gain physical access to a device if it is stolen. Attackers have the ability to tamper with individual components of a device, which can result in the device functioning incorrectly.

5. *Attacks that utilize different kinds of software*: These kinds of attacks are possible whenever there are security issues in em-

bedded system software or device firmware, as well as whenever there are vulnerabilities in cloud servers or backend applications.

10.11   THE MOST EFFECTIVE STRATEGIES FOR ENSURING THE SAFETY OF IoT SYSTEMS

The three most important aspects of an IoT system are the devices, the networks, and the data; following the best practices for IoT security can help you safeguard all three of these aspects more effectively. Let's begin by discussing the best ways to safeguard our technological gizmos.

1.   Safe mobile devices:
   - Verify that the hardware cannot be altered in any way. Intruders may steal IoT devices in order to access private data or to tamper with the devices themselves. To ensure that the data on the device is kept secure at all times, check that your product cannot be altered. You can ensure the device's physical security by taking precautions such as installing port locks, camera covers, robust passwords for the boot level, and other safeguards that, in the case of tampering, will render the device useless.
   - Make available software patches and updates. Continuous device upkeep requires additional expenses. Nevertheless, installing patches and updates on a regular basis is the only method to effectively assure product security. It is highly recommended to adopt security upgrades that are both mandatory and automatic, and do not involve any action on the part of end users. Customers need to be informed of the amount of time remaining in the product's support period as well as the next steps to take once it has ended. After you have published your system to the public, it is imperative that you keep an eye out for potential security flaws and produce patches as required.
   - One must perform exhaustive testing. Penetration testing should be your primary tool for locating vulnerabilities in the firmware and software of IoT devices and for reducing

the attack surface to the maximum extent possible. Static code analysis is able to locate even the most egregious of flaws, and dynamic testing is able to unearth vulnerabilities that have been carefully camouflaged.

- Implement data protections for devices. Devices connected to the IoT need to ensure data security before, during, and after use. Make sure that the cryptographic keys are stored in a memory that is not volatile on the device. You might also make it possible for consumers to dispose of used products in a way that does not involve the disclosure of personal information.

- Ensure compliance with the performance specifications for the component. In order to provide the best possible performance, hardware for IoT devices needs to comply with a predetermined set of performance standards. For example, IoT devices ought to have high computational capability while consuming less electricity. Devices also need to ensure secure and dependable wireless connectivity, as well as permission and data encryption. In an ideal world, your IoT system should continue to operate normally even if its internet connection is temporarily severed.

2. Reliable networks:
  - Ensure that the authentication is done in a secure manner. It is feasible to accomplish this by utilizing separate default credentials. If you want to ensure that your products will continue to be useful in the future, you should identify or address them using the most recent protocols. If it's at all practicable, incorporate MFA into your product.

  - Enable encrypted means of secure communication and other such methods. The communication between devices also needs some form of security protection. However, in order to account for the limited capabilities of IoT devices, cryptographic methods will need to be adjusted. Lightweight cryptography and Transport Layer Security are two options for securing these respectively. When designing an IoT architecture, you have the option of utilizing either wireless or wired technologies, such as RFID, Bluetooth, cellular,

ZigBee, Z-Wave, Thread, and Ethernet. Additionally, one is able to ensure the safety of one's network by utilizing optimized protocols like IPsec and Secure Sockets Layer.

- Reduce the bandwidth used by the device. Allow only the minimum amount of network traffic necessary for the IoT device to perform its intended function. It is advised that the device be programmed to restrict bandwidth at the hardware and kernel level and to flag activities that may be suspect. The service will be protected from any potential DoS attacks if this is done. Because malware can be used to take control of the device and use it as part of a botnet to launch distributed DoS attacks, the product should also be engineered to reboot and clean code if malware is identified. This is because malware can be used to take control of the device.

- Develop networks that are divided into segments. The protection of a next-generation firewall can be implemented by first separating large networks into a large number of smaller networks. For this purpose, you might make use of IP address ranges or VLANs. For secure access to the internet, you need to incorporate a virtual private network (VPN) into your IoT system.

3. Confidential information:

- Be sure to protect any sensitive information. Install unique default passwords for each product or insist that users update their passwords immediately upon the device starting up. Use a method that is capable of providing a high level of authentication to ensure that only authorized users can access the data. You can also choose to give a reset mechanism, which will enable the erasure of private data as well as the wiping of configuration settings in the event that the user decides to return or resell the device. The protection of each individual's privacy will be significantly strengthened as a result of this.

- Only collect the absolutely necessary information. Make sure that your IoT device is only collecting the information that is necessary for it to perform its functions. This will reduce the risk of data leakage, preserve the privacy of customers, and take care of any potential problems that could arise from failing to comply with a variety of data protection standards, rules, and regulations.

- Discussions should be protected that take place via networks. For the sake of greater security, try to restrict any unwanted communication that occurs between your product and the IoT network. If you want to keep your communication secure, you shouldn't rely just on the network firewall; instead, you should make your product invisible to incoming connections by default. Use encryption techniques such as the Advanced Encryption Standard, Triple Data Encryption Standard (DES), Rivest-Sharmir-Adelman (RSA), and Digital Signature Algorithm that are adapted to the requirements of IoT devices.

10.12   CONCLUSION

It's vital to start considering security at the research and development phases of IoT initiatives. However, ensuring the effective cybersecurity of devices, networks, and data in IoT contexts is difficult due to the frequency of incursions and the complexity of checking for potential system vulnerabilities. Applications that use the IoT can make it difficult to apply rigorous security safeguards. Adding security measures could potentially raise the price of a product as well as extend the amount of time it takes to produce it, both of which are undesirable for businesses. Hardware limits are another potential issue.

References

[1]   Bronevetsky, G., I. Laguna, B. R. de Supinski, and S. Bagchi, "Automatic Fault Characterization via Abnormality-Enhanced Classification," in *Proc. IEEE/IFIP Int. Conf. Depend. Syst. Netw. (DSN)*, 2012, pp. 1–12.

[2]   Bychkovskiy, V., S. Megerian, D. Estrin, and M. Potkonjak, "A Collaborative Approach to In-Place Sensor Calibration," in *Information Processing in Sensor Networks*, Heidelberg, Germany: Springer, 2003, pp. 301–316.

[3]   Balzano, L., and R. Nowak, "Blind Calibration of Sensor Networks," in *Proc. 6th Int. Conf. Inf. Process. Sensor Netw.*, 2007, pp. 79–88.

[4]   Neumayer, S., and E. Modiano, "Network Reliability with Geographically Correlated Failures," in *Proc. IEEE INFOCOM*, 2010, pp. 1–9.

[5] Sharma, A. B., L. Golubchik, and R. Govindan, "Sensor Faults: Detection Methods and Prevalence in Real-World Datasets," *ACM Trans. Sensor Netw.*, Vol. 6, No. 3, 2010, pp. 1–39.

[6] Jeffery, S. R., G. Alonso, M. J. Franklin, W. Hong, and J. Widom, "Declarative Support for Sensor Data Cleaning," in *Proc. Int. Conf. Pervasive Comput.*, 2006, pp. 83–100.

[7] Tancreti, M., M. S. Hossain, S. Bagchi, and V. Raghunathan, "AVEKSHA: A Hardware–Software Approach for Non-Intrusive Tracing and Profiling of Wireless Embedded Systems," in *Proc. 9th ACM Conf. Embedded Netw. Sensor Syst. (Sensys)*, 2011, pp. 288–301.

[8] Gross, K. C., K. Baclawski, E. S. Chan, D. Gawlick, A. Ghoneimy, and Z. H. Liu, "A Supervisory Control Loop with Prognostics for Human-in-the-Loop Decision Support and Control Applications," in *Proc. IEEE Conf. Cogn. Comput. Aspects Situation Manag. (CogSIMA)*, 2017, pp. 1–7.

[9] Guo, B., D. Zhang, Z. Wang, Z. Yu, and X. Zhou, "Opportunistic IoT: Exploring the Harmonious Interaction Between Human and the Internet of Things," *J. Netw. Comput. Appl.*, Vol. 36, Nov. 2013, pp. 1531–1539.

[10] Szewczyk, R., J. Polastre, A. Mainwaring, and D. Culler, "Lessons From a Sensor Network Expedition," in *Proc. Eur. Workshop Wireless Sensor Netw.*, 2004, pp. 307–322.

[11] Dong, X. L., L. Berti-Equille, and D. Srivastava, "Integrating Conflicting Data: The Role of Source Dependence," *Proc. VLDB Endown.*, Vol. 2, No. 1, 2009, pp. 550–561.

[12] Karimi, M., and H. Kim, "Energy Scheduling for Task Execution on Intermittently-Powered Devices," in *Proc. 9th Embedded Oper. Syst. Workshop*, 2019, pp. 1–6.

[13] Maeng, K., and B. Lucia, "Adaptive Low-Overhead Scheduling for Periodic and Reactive Intermittent Execution," in *Proc. 41st ACM SIGPLAN Conf. Program. Lang. Design Implement. (PLDI)*, 2020, pp. 1005–1021.

[14] Cranor, L. F., "A Framework for Reasoning About the Human in the Loop," in *Proc. UPSEC*, 2008, pp. 1–15.

[15] IoT Security: The Top 10 Threats and How to Protect Yourself by Symmetry Electronics.

[16] "The Security Risks of IoT Devices," ZDNet.

[17] "The Most Common Types of IoT Malware," Zscaler.

[18] "The Most Dangerous IoT Malware and How to Protect Yourself," Microsoft.

[19] "BotenaGo Botnet Targets IoT Devices with Over 30 Vulnerabilities," Trend Micro, Tokyo, Japan.

11

FUTURE SCOPE AND CONCLUSION

Mobile networks have been in operation for the past four decades, during which time a new generation of mobile networks has emerged at least once every 10 years. The mobile communication sector has flourished thanks to the development of various technologies in the decades following the establishment of the first commercial mobile network in 1980. The cellular technologies that make it possible for new generations of mobile networks to exist each have their own unique requirements, and those requirements are met by the cellular technologies. Advanced Mobile Phone Service (AMPS), General Packet Radio Service (GSM), Universal Mobile Telecommunication System (UMTS), CDMA2000, and LTE are all examples of cellular technologies. In this book, we discussed 5G smart antennas (SAs) and their applications for the IoT. The most recent version of cellular technology, referred to as 5G, is made feasible by the NR technology. This technology is based on OFDM access (OFDMA). The high efficiency of 5G and its capacity to support a large number of devices make it possible for numerous industries to undergo modernization. In addition to this, it is possible for it to function in a number of other frequency ranges, including both high and low frequencies. The higher-frequency bands of the 5G network have limited penetration; nevertheless, because the 5G higher-frequency bands have extremely low latency (less than 1 millisecond), they are excellent for providing real-time services. Researchers are interested in the three primary drivers of 5G application scenarios, which are eMBB, mMTC, and uRLLC, and they could be able to assist you in learning more about these three essential foundations of 5G technology. The typical downlink speed in 5G was approximately 150 Mbps, despite the fact that it is possible for 5G to achieve rates of over 10 Gbps. The vast majority of 5G installations are nonstand-alone (NSA) at this point, which indicates that they are not complete 5G deployments. The technology that results from the combination of 4G and 5G networks is referred to as 5G NSA access. Broader channels (speed), lower latency (responsiveness), and higher bandwidth are the three new qualities that the fifth generation of wireless technology brings to the table (the ability to connect a lot more devices at once).

IoT and 5G were practically designed for one another, and they round out and complete one another in the most desirable ways. As sensors become more integrated into the activities we participate in on a daily basis, the skeleton evolves into the gadget, and 5G gives such gadgets the ability to function. The 2G, 3G, and 4G network standards are difficult to work with these gadgets. You can get started with 4G, but 5G really brings out their beauty. Antennas take on a challenging function within the dynamics of the IoT. Fiber-optic cable paved the way for the development of this technology by making possible the utilization of high-speed 5G networks that featured a variety of wireless channels and reliable delivery. The use of cloud-like concepts to both radio access networks and core networks results in improved connectivity dynamics. Cloud-radio access network (C-RAN) is an example of a rapidly expanding worldwide cloud radio access network that places an emphasis on the pooling of baseband units (BBUs) and the utilization of cloud technologies. Despite the fact that they are two distinct forms of technology, they both have the following characteristic in common: Both of these require connectivity that is reliable and long-lasting, operating at high speeds, and carrying a significant amount of data. Real-time cloud visualizations, which included real-time warnings and photos, were utilized in order to monitor and run the plant from a significant distance. This was made possible by antennas for LPWAN.

As personal electronics and cell phones have become more sophisticated over the past few years, product engineers have developed methods that almost eliminate the need for antennas on the vast majority of these devices, particularly smartphones. Consider the following: When was the last time you needed to modify the antenna on your iPad or Samsung phone in order to increase the device's reception? The millennials, teenagers, and young children of today are ignorant of the major role that the antenna played in the invention of the radio, television, and mobile phones; nonetheless, this information is readily available. Previous generations were aware of the necessary steps to modify an antenna in order to achieve better reception.

Despite this, antennas are still often used. In fact, scientists and engineers are currently working on developing new antenna technologies for the future, in addition to inventive ways to combine antennas into technologies that have already been proven effective.

11.1   ANTENNAS 3-D PRINTING TECHNOLOGY

There is a strong probability that you are familiar with the ways in which 3-D printing is revolutionizing the industrial industry. Businesses in every region of the world are scratching their heads to discover how they can give customers the ability to print whatever they want, whenever they want it. It is fascinating to see that the antenna sector has embraced the use of 3-D printing.

One of the ways that 3-D printing is influencing the future of antennas is the way that product engineers are increasingly creating traditionally shaped antennas that are both lighter in weight and less expensive to produce than what is possible with the manufacturing methods that are currently in use. The fact that these 3-D-printed antennas are often more powerful than traditional antennas is another reason why they are notable. Optomec's Aerosol Jet® is capable of 3-D printing complex electrical structures and components, such as antennas, for a variety of applications, including consumer electronics, wearables, IoT, aerospace, and defense.

11.2   NEW TREND TO GET ANTENNAS THAT ARE MUCH SMALLER: MINIATURIZATION

Product engineers are exploring creative alternatives to 3-D printing in order to find new ways to decrease the size of antennas. The process of making antennas smaller is referred to as antenna miniaturization, and it will surely play an important part in the future development of technology related to antennas.

Scientists are currently working on developing antennas that are between a tenth and a hundredth of the size of those that are currently on the market. These new ultrasmall antennas are distinct from traditional antennas in that, rather than relying on electromagnetic waves for communication, they use radio waves instead. According to the research that was published in *Nature* [1], the authors' miniantennas were capable of transmitting and receiving signals at a frequency of 2.5 GHz—almost 100,000 times more effectively than a conventional ring antenna.

Each year, an imaginative blogger in the field of technology writes a wordy article in which they assert that an innovative technology that does not require antennas would soon do away with the need for mobile phone towers. However, the fact remains that, as a result of the never-ending need for increased bandwidth, mobile phone service providers are currently making more money off of the sale of bandwidth than they are from the sale of actual phones. In order for carriers to satisfy this need for bandwidth, they will need to maintain their reliance on the existing network of cell phone towers.

Cell phone towers will continue to play a significant role in mobile communication until companies such as AT&T and Verizon discover a way to produce more bandwidth than they can sell, which is a problem that these companies do not want to solve. To put it another way, cell phone towers are going to be around for an extremely extended period of time. Although cell phone towers will continue to play a role in the development of mobile phone technology, 5G mobile technology is currently on the horizon and is expected to revolutionize the mobile phone business when it becomes available. The most significant and interesting innovation in antenna technology that will be associated with the development of 5G will be the introduction of more compact cell towers.

Due to the fact that these experimental cell phone towers are so small, mobile phone carriers are able to attach them to preexisting structures such as light poles and rooftops. Because they are so much smaller than the present mobile phone towers, the phone companies should be able to install these tiny antenna towers in rural areas across the country. The greater the number of antenna towers that are there, the better your cell phone service will be. The experts believe that the 5G technology will improve cell phone connectivity since there will be less dropped calls and better overall reception. This is due to the fact that there will be so many nearby small antenna cell towers.

11.3 FUTURE ANTENNA TECHNOLOGY

In spite of the fact that antennas are generally considered to be an outmoded kind of technology, telecommunications businesses are attempting to innovate the antenna in preparation for the next wave of electronic devices. Product engineers are currently utilizing 3-D printing or other advanced manufacturing techniques to make antennas that are more powerful than ever before, as well as smaller, lighter, and more powerful than they have ever been before. The rollout of 5G technology by mobile phone operators is planned, and it will usher in a return to the use of much smaller cell towers. Despite these developments, traditional mobile phone towers and antennas of a greater size will continue to play an essential role in deciding the course that human communication will take in the future.

The speed of the 5G network is not solely attributable to the frequency of the signal, but also to new technological advancements in the antennas. On the other side, 5G will be further away from the 4G towers and will be more susceptible to interference from high-frequency obstructions such as tall buildings and trees. As a consequence of this, several 5G towers will need to be installed for even coverage in order to provide the required level of speed and service. This will need a significant investment of both money and time. The massive MIMO (mMIMO) approach that has been suggested is able to successfully cut down on interference at the edge server (ES) while simultaneously preserving the functionality of the 5G IoT system in mmWave. It will be capable of handling needs that are exceptionally high in terms of bandwidth, integrity, and latency. This research also investigates recent developments in resource constraint agents (RCAs), analytical methods, and other aspects that make them appropriate for application in 5G communication networks. mMIMO systems make use of a large number of antennas to compensate for the effects of distortion, blurring, and entanglement. The multitude of antennas that are utilized in huge MIMO makes the system more complicated and increases the costs of the hardware. The development of mMIMO technology must make use of technologies that are both economical and small in order to lower the processing complexity rather than the equipment bulk. Low-cost hardware may exacerbate technological problems such as nonlinearities, magnetization noise, amplifier distortions, and even conceptual mismatch. Utilizing the appropriate compensating measures allows for the impact of the equipment constraint to be reduced; nevertheless, this does not mean that it can be eliminated entirely.

11.4   CONCLUSION

It is conceivable for 5G networks to achieve speeds of up to 20 Gbps, making them potentially one hundred times quicker than 4G and 4G LTE networks. As a result of reductions in latency, new applications such as the IoT and AI will be able to support real-time communication. As a result of the acceleration of IoT adoption brought about by 5G, business connection will expand to an unprecedented level. Electronic eMBB, URLLC, and enormous machine-type connections will be needed for AAS or small antennas. These three applications are considered to be the three pillars of early 5G use cases. The usage of multiple antenna techniques, such as mMIMO, sub-6 GHz, including mmWave carrier aggregation, and full-dimensional (FD) directional antennas, are examples of approaches that facilitate the adoption of 5G. In order to achieve the exceptional outcomes targets that have been set, the static antenna technology that is currently in use must be replaced with dynamically operated density patch antennas that are analogous to the staggered array antennas that are utilized in advanced military AESA technologies. Antenna architectures rely on dynamic MIMO networks that are fully interconnected and tightly packed. Over the course of the past several years, researchers have made a large number of attempts, with a wide variety of ideas and methods, to address the issue of the 5G antenna for the IoT application.

Research on the IoT and 5G wireless networks can assist developing countries and the entire world in providing better social services. Some of the areas of focus for 5G research include MIMO and mmWave communication technology, as well as network security, the management of data traffic, the development of cloud algorithms, and other network topics.

Reference

[1]   Martin, B., "Ultra-Small Antennas Point Way to Miniature Brain Implants," *Nature*, Vol. 548, No. 7668, August 23, 2017, pp. 503–504, doi:10.1038/nature.2017.22507.

ABOUT THE AUTHOR

Prutha P. Kulkarni is an accomplished assistant professor at the Vishwakarma Institute of Information Technology in Pune, Maharashtra, India. With a specialization in antenna miniaturization, she possesses extensive research experience in the field of antennas and metamaterials for IoT applications. Her current research work primarily revolves around the design of metamaterial-inspired antennas, miniaturized antennas for GPS, CubeSat, and IoT applications, as well as her keen interest in 5G technology.

With Australian and Indian patents to her credit, she is also an accomplished author of technical books on wireless communications, has awards, and has published numerous papers in well-regarded journals and international conferences. Beyond her academic achievements, Dr. Kulkarni actively contributes to the academic community as a respected reviewer for prestigious journals and conferences like IEEE. She is a proud member of the IEEE Antennas and Propagation Society (AP-S) and Microwave Theory and Technology Society (MTT-S), where she currently serves as the vice-chair of the IEEE APS/MTT/EMC joint chapter in the Pune section.

Dr. Kulkarni has completed her PhD in information and communication engineering, specializing in the miniaturization of LP and CP antennas, from Anna University in Chennai, India in 2023. She completed her ME in electronics and telecommunication from Pune

Institute of Computer Technology in 2013, and her BE in electronics and telecommunication from Vishwakarma Institute of Information Technology in Pune, Maharashtra, India in 2008.

INDEX

Ultrawideband Antennas for Microwave Imaging Systems,
Tayeb A. Denidni and Gijo Augustin

Ultrawideband Short-Pulse Radio Systems, V. I. Koshelev,
Yu. I. Buyanov, and V. P. Belichenko

*Waveguide Components for Antenna Feed Systems: Theory and
CAD,* Jaroslaw Uher, Jens Bornemann, and Uwe Rosenberg

Wearable Antennas and Electronics, Asimina Kiourti and
John L. Volakis, editors

For further information on these and other Artech House titles, including previously considered out-of-print books now available through
our In-Print-Forever® (IPF®) program, contact:

Artech House	Artech House
685 Canton Street	16 Sussex Street
Norwood, MA 02062	London SW1V HRW UK
Phone: 781-769-9750	Phone: +44 (0)20 7596-8750
Fax: 781-769-6334	Fax: +44 (0)20 7630 0166
e-mail: artech@artechhouse.com	e-mail: artech-uk@artechhouse.com

Find us on the World Wide Web at: www.artechhouse.com